TREES
OF THE ROYAL PARKS

A GUIDE TO COMMON
AND RARE SPECIES FOUND IN THE
ROYAL PARKS OF LONDON

THE
ROYAL
PARKS

© Crown Copyright 1996

First published in the UK in 1996 by The Royal Parks
The Old Police House, Hyde Park, London W2 2UH

Applications for reproduction should be made
to The Royal Parks

British Cataloguing-in-Publication Data
A catalogue record for this book is available from
the British Library
ISBN 1-898506-08-6

Printed by the KPC Group

Contents

7
Introduction

15
Hyde Park

23
Kensington Gardens

35
St James's Park and Green Park

45
Regent's Park

55
Greenwich Park

65
Richmond Park

77
Bushy Park

88
Envoi

89
Glossary

90
Bibliography

91
Index of Trees in the Royal Parks

Text
J.G. Berry

Photographs
Martin Jones

Editor
Maggie O'Hanlon

Design and maps
Penny Jones

The Royal Parks are grateful to
J.O. Hambro & Company for their generous
contribution towards the cost of planting
and labelling trees in St James's Park
and Green Park and to the production
of this publication.

If you have enjoyed the beauty of The Royal Parks
and would like to contribute towards planting
new trees, please send donations to:
The Prince of Wales Royal Parks Tree Appeal
(Registered Charity No. 802610)
8th Floor, 4 Golden Square, London W1R 3AE
Telephone 0171 734 8312

INTRODUCTION

INTRODUCTION

ANY MAP of London will show that the inner Royal Parks – Kensington Gardens, Hyde Park, Green Park, St James's Park and Regent's Park – fit very neatly into the central conurbation. The first four linked, open spaces, richly furnished with trees, connect the shopping areas and museums of Kensington, Buckingham Palace – the seat of the ruling monarch, the centres of Britain's diplomacy and government, the River Thames, and yet more shopping areas and museums, in a wonderfully picturesque manner. Such is the abundance of delightful vistas, sufficient to satisfy even the most ambitious camera- or camcorder-toting tourist, that it might be thought that this had all been designed by a super-efficient and inspired planner. Then, at increasing distances from the centre, are the other three Royal Parks – Richmond Park, Greenwich Park and Bushy Park, filling in the gaps in London's necklace of large commons and forests.

And yet all these fabulous green properties in London and its environs, which relieve the oppressive atmosphere of bricks, concrete and tarmac, thunderous with traffic, are happily accidental. The Royal Parks were nearly all acquired as hunting-grounds and, in spite of their long history, they boast few really ancient trees – no 1,000-year-old veteran oaks. The Parks have endured too many ups and downs to retain many ancient trees and woodland, and there have been too many storms and fashions in landscaping for these essentially long-living plants to attain vast size and age. The principal problem for trees of any quality is that their financial value increases with their girth, and many impoverished owners must have been tempted to cut down and sell

Facing page:
Sweet Chestnut
Castanea sativa
Greenwich Park

Right: Persian Ironwood
Parrotia persica
Bushy Park
Woodland Garden

INTRODUCTION

Below: Oriental Bittersweet
Celastrus orbiculatus
Bushy Park
Woodland Garden

Above: Cypress Oaks
Quercus robur 'Fastigiata'
Hyde Park

their trees for timber. Nevertheless, there are a few trees of very respectable antiquity in the Parks, such as John Martin's Oak in Richmond Park (probably 500+ years old) and the picturesque Sweet Chestnut ruins in Greenwich Park (about 350 years old).

The seventeenth century has left behind a record of contemporary fashions in tree-planting in most of the Parks, although not all show the formal development of *patte d'oie* and radiating avenues that can be seen to such excellent effect in Greenwich Park or Kensington Gardens. Bushy Park has one absolutely splendid Chestnut Avenue, and a few minor avenues, while the avenues in Hyde Park's Broad Walk and Rotten Row are crooked.

The eighteenth- and early nineteenth-century fashion for the 'landscape garden' and sophisticated irregularity had its influence on the planning and planting of the Parks. This, however, was not as great as one might expect, bearing in mind that the movement's most powerful figure, Lancelot ('Capability') Brown, Master Gardener to His Majesty George III from 1764 until 1783, was responsible for the management of Hampton Court, St James's Park and Richmond Park. During this time, he prepared an undated scheme for deregularising the landscape of St James's Park – a scheme remarkably similar to that of John Nash in 1827. Capability's scheme was never carried out but John Nash's was. All the Royal Parks contain greater or lesser fragments of formal planting schemes to this day, although none have been given the fully 'irregular' treatment. Even Richmond Park has the absolutely straight Queen's Ride, cut through the woodland between White Lodge and the Richmond Gate, together with some short fragments of avenue, and St James's Park has its picturesque and irregular nineteenth-century design straitjacketed inside the formal avenues and lines of a layout surviving from two centuries before.

Not all the Parks were acquired for hunting (in its medieval sense, a 'park' was an enclosed piece of land for beasts of the chase, not necessarily containing trees). Kensington Gardens, from the first, was a garden rather than a park and, together with the building which was to become Kensington Palace, was bought from Sir Daniel Finch by King William III. Driven by the arduous duties to both his adopted country and his native Holland, and martyred by asthma, King William wanted a peaceful place where he could escape the fumes (largely from coal fires and a polluted river) and stresses of central London. These Gardens remain the most complete surviving, large seventeenth-/early eighteenth-century country-house garden in central London.

During the nineteenth century there was some return to formal planting. It is difficult to achieve any measure of pomp and circumstance without arranging trees in formal lines and phalanxes, and this accounts for the very formal spine – the Broad Walk – in Regent's Park, the only London Royal Park designed and laid out wholly in the nineteenth century. This was to have led to a small *pied à terre* (the Guingette) for the Prince Regent.

Above: Ginkgo
Ginkgo biloba
Greenwich Park
Flower Garden

Right: Red Oak
Quercus rubra
Kensington Gardens

Trees of The Royal Parks

In spite of the Parks' different histories, or sometimes because of them, it is the trees that dominate their character. The London Plane (*Platanus* x *acerifolia*), because of its diffuse crown lifted high above the dusty world, provides the matrix within which other landscape elements are set, in St James's and Green Parks, and in large parts of Hyde Park and smaller parts of Kensington Gardens. The stately lime avenues have the ascendancy over other plantings in Kensington Gardens, whereas the Common Oak (*Quercus robur*) provides the dominant colour, and the robust contrast to the vast green expanses, in Richmond and Bushy Parks. The oldest Park – Greenwich – has an atmosphere of antiquity in its buildings that is enhanced by the eccentric hulks of its ancient Sweet Chestnut trees. Regent's Park is more difficult to encapsulate – its basic character might yet emerge during the course of the next few centuries.

London Plane

The origins of the London Plane are vague. Academics discussing its provenance use the kinds of words we fear: 'may', 'probably', 'possibly', 'it has been said', 'reliable authorities have stated'! This tree has been around since the seventeenth century and exhibits, in varying proportions, the characteristics of two very different species: the confusingly named Oriental Plane (*Platanus orientalis*), which is actually a European species, and the Western

Facing page: Plane trees
Platanus spp.
St James's Park
Birdcage Walk

Right: Paper-bark Birch
Betula papyrifera
Greenwich Park

Plane (*Platanus occidentalis*), an American species. The first is a fine sturdy tree with deeply lobed leaves, resistant to disease and wind, and capable of long life and grand dimensions in an English climate; there are a few in the Royal Parks. This species seems to have crossed with the Western Plane, perhaps more than once, somewhere in southern Europe shortly after its introduction from the USA. Early this century, W.J. Bean, a giant of tree lore, stated that, at Kew, Western Planes raised from seed would grow a few feet in height in a hot year and then fall over and die, suffering from the vegetable equivalent of the common cold.

As a result of this hybridisation, a London Plane – normally quite happy and cheerful in the cities of southern Britain (it is a different story in northern Britain) – will vary in its characteristics according to whether it is related more closely to the hardy Oriental Plane or to the less resistant Western Plane. Not only are the leaves and bark patterns infinitely variable between individual trees but the Park Managers, who see these trees every day, have noted that trees which are related more closely to the Western Plane show a greater vulnerability to salt pollution and leaf anthracnose (a serious fungus disease) than superficially similar trees showing affinities with the Oriental

Plane. Recognising these relationships in the field, however, is more difficult than describing them on paper.

A very interesting variant of the London Plane, which is larger in all its parts and seems to be much more vigorous, occurs in both Green Park and St James's Park. This is the cultivar *Platanus* x *acerifolia* 'Augustine Henry', named after a very distinguished arborist of the late nineteenth and early twentieth centuries.

WILLOWS

A similar situation exists among the willows of the Royal Parks. The weeping willows are the most important in terms of both numbers and landscape effect. The taxonomy of the two commonest – the Golden Weeping Willow (*Salix* 'Chrysocoma') and the Babylon Willow (*Salix* x *sepulcralis*) has a long history of academic argument and controversy, although the common names have been retained for the purposes of this book.

Current thinking considers the Babylon Willow to be a tender species, all but extinct in Britain, which has been replaced by an assembly of hybrids of indeterminate ancestry, gathered together for convenience under the name *Salix* x *sepulcralis*. This includes both the Golden Weeping Willow, with long twigs, beautiful yellow bark and branches weeping to the ground, and the Babylon Willow, with short, green or brown twigs distributed more informally about the crown in pendulous masses. Visitors will notice a few Babylon Willows at the watersides which have a more silvery appearance than usual.

ORIGINS OF THE TREES

Judging from the trees present in the Parks, many were obtained from the same source. All the Parks have specimens of Schwedler's Maple (*Acer platanoides* 'Schwedleri'), Silver Maple (*Acer saccharinum*) – usually its cultivar 'Wieri' – and Southern Magnolia (*Magnolia grandiflora*) in which the individuals have the characteristics of seed-raised plants, i.e. variability. The waterside willows probably have a similar origin; there are both kinds of weeping willow, White Willow (*Salix alba*), Crack Willow (*Salix fragilis*) and, occasionally, Silver Willow (*Salix alba* f. *argentea*). Lombardy Poplar (*Populus nigra* 'Italica'), a diversity of Black Poplars (*Populus nigra*), alder (*Alnus* spp.), Swamp Cypress (*Taxodium distichum*) and, sometimes, Dawn Redwood (*Metasequoia glyptostroboides*) can also be found.

How to Use This Book

Visitors to the Royal Parks need not feel ashamed if they are unable to identify every tree, even with the aid of the most authoritative identification guide. Trees, like other living things, are very variable and some, such as species of oaks, birch, poplars and willows, readily produce hybrids. Nurserymen, also, are constantly producing new named variants, such as the proliferation of white-stemmed birches, with a multiplicity of cultivar names.

The common names used in the book are those in most popular use and will generally be found in the pocket identification guide compiled by Alan Mitchell (1982). In addition, the Park Managers are making every effort to ensure that all the trees mentioned in the walks are accurately and securely labelled. In every chapter, Latin names have been included at least once to enable precise identification. Remember that there will be more than a hundred species of tree in any one Park, of which only a very small number can be named or described in the tree walks. Those not mentioned in the text will be the more common species.

Medlar
Mespilus germanica
Kensington Gardens
North Flower Walk

Hyde Park

14

HYDE PARK

F EW PEOPLE visit Hyde Park solely to enjoy its trees and it has not always existed in its current well-planted state. For a very long time it has been used heavily by the general public, and the Parade Ground in particular has always been the venue for military displays and all kinds of shows and public meetings. It is not an arboretum in the true sense of the word, although a great part of its attraction lies in the combination of tree masses and single trees that frame the succession of variously shaped green spaces leading to the shining water of the Serpentine. Nevertheless, in a survey carried out in 1982, no less than 159 distinct species of trees were found. Since then, in spite of the inevitable losses resulting from storms and normal dilapidation, planting has continued and the total number of species is now far greater.

Like all the other Royal Parks, Hyde Park was not established on first-class agricultural soil. Such a rich loam might have been fine soil for trees but it would have encouraged over-vigorous ground-covering plants, which would have grown too tall for a deer park and made hunting impossible. Soil has always been a problem for Park Managers. Near the original course of the Tyburn Stream is every kind of alluvium but, in other places, the soil is dry and poor. Charles Lamb wrote to Wordsworth early in the nineteenth century: 'The whole surface of Hyde Park is dry crumbling sand. Not a vestige or a hint of grass ever grows there.'

Facing page: Plane trees
Platanus spp.
The Broad Walk

Right: Black Mulberry
Morus nigra
Hybrid Wingnut
Pterocarya x *rehderiana*
Hyde Park Corner Lodge

In 1960, a slice of the tree belt and the East Carriage Drive were acquired by local authorities to ease traffic congestion in Park Lane and, at the same time, a vast underground car park was constructed beneath the Parade Ground. This cost the Park further losses of mature trees and left severe drainage problems because of the way in which the soil was re-instated on top of the concrete roof. Indeed, there are now large areas of the Park where potentially large trees may not be planted because of the effects that they might have on the roof of the car park.

The great storms during the 1980s led to a reduction in the number of screen trees and it has been difficult to replace these quickly. However, great efforts have been made to fill the gaps in the tree cover in the northeastern quarter of the Park.

The most famous tree in the Park is the Reformers' Tree. This stood at the junction of five major pathways and was a meeting place for rejoicers and complainers from the early nineteenth century onwards. Meetings so often developed into near riots, actual riots, or something close to a pitched battle that, in 1872, the Commissioners of Woods and Forests set up Speakers' Corner on the fringe of the Park; even this has not put a stop to occasional disturbances on the Parade Ground. The original tree is long defunct but its latest replacement, a Red Oak (*Quercus rubra*), was planted in the same spot in the early 1970s by Lord Callaghan.

A Tree Walk

Enter the Park via the **Apsley Gate** through the Ionic Screen, the work of Decimus Burton. Before crossing the **South Carriage Drive**, inspect the back garden of **Hyde Park Corner Lodge**, now the Visitor Information Centre, where there is an old Black Mulberry Tree (*Morus nigra*) growing in the shade of a picturesquely branched Hybrid Wingnut (*Pterocarya x rehderiana*). Cross the Drive near the **Queen Elizabeth Gate** and approach the **Broad Walk** – a magnificent avenue of London Planes with an average height of 30m (100ft). Now cross to the nearest planted borders, an area full of fine plants. Standing just in the open parkland, a few paces away from the eastern fence, is a Kentucky Yellow Wood (*Cladrastis lutea*), so called because its wood turns yellow after conversion into timber. At the rear of the nearest border is a Ginkgo (*Ginkgo biloba*).

The border on the other side of **Serpentine Road**, opposite the **Cavalry Memorial**, has a cluster of interesting species, including a small *Pittosporum tenuifolium*, another Ginkgo hidden alongside in the shrubbery and a fine large Highclere Holly (*Ilex x altaclerensis*). The canopy is provided by a Horse-chestnut and a Lucombe Oak (*Quercus x hispanica* 'Lucombeana'). In the lawn alongside is a Cut-leaved Silver Maple (*Acer saccharinum* 'Wieri'), accompanied by a very rare Caucasian Elm (*Zelkova carpinifolia*), with its unforgettable brush-shaped crown. Many more specimens of Cut-leaved Silver Maple can be found in various places in the Park.

From here, walk towards the **Dell** and the **Ornamental Gardens**. Near the northern side of the cross-path is a young Keaki (*Zelkova serrata*), a Japanese relation of the elm, behind which are three fine Cypress Oaks (*Quercus robur* 'Fastigiata'); these are still quite young and can be expected to attain a great

Facing page: Ash trees
Fraxinus spp.

Right: Weeping Beech
Fagus sylvatica 'Pendula'

height eventually. The small specimen tree with attractively crooked branches, growing on a slight rise north of the central path, is a Cornelian Cherry (*Cornus mas*). In late winter, it bears myriads of tiny yellow flowers on its otherwise naked branches and, in autumn, cherry-sized fruits closely resembling cornelian beads. In this large area are several Weeping Beeches (*Fagus sylvatica* 'Pendula'), the best of which is in the fenced enclosure of the Dell.

Walk towards the single Cypress Oak and the **Holocaust Memorial**. In early summer, it is worth making a detour to the Serpentine Road to see the group of Pink-flowered False Acacias (*Robinia* x *ambigua* 'Decaisneana') in flower. To the right, on the corner opposite the Dell, is a Copper Beech (*Fagus sylvatica* 'Purpurea'), and a Caucasian Wingnut (*Pterocarya fraxinifolia*), and around the Holocaust Memorial are some common Silver Birches (*Betula pendula*), well established among the ground-cover plants. A number of exotic birch species have recently been planted among and around them to enhance the landscape effect and botanical interest. Walk around the attractively planted Dell, past the Caucasian Wingnut. In addition to the Weeping Beech already mentioned, you will find splendid Highclere Hollies, a delicately beautiful Japanese Maple (*Acer* spp.) a vigorously growing Dawn Redwood (*Metasequoia glyptostroboides*), a Common Magnolia (*Magnolia* x *soulangeana*) and a variety of choice shrubs. The white-stemmed birches are mainly Himalayan Birch (*Betula utilis*) while the very exotic, pinky-purple-stemmed

Caucasian Wingnut
Pterocarya fraxinifolia
Rotten Row

Black Walnut
Juglans nigra

saplings on the upper side of the Dell, not far from the Paper-bark Maple (*Acer griseum*), are Red-barked Birch (*Betula albosinensis* f. *septentrionalis*).

Make a short detour from here to look at the tree lines of **Rotten Row**. Although lacking the uniformity of the Broad Walk, the Row is an impressive avenue of planes, limes and horse-chestnuts.

It is possible to make a circumnavigation of the **Serpentine** by walking on the Lido side. Here you will find White Willow (*Salix alba*), mature Black Walnut (*Juglans nigra*) and sapling Common Walnut (*J. regia*), a mature Indian Bean Tree (*Catalpa bignonioides*) near the **Lido** building, and a recently planted group of Japanese Golden Elm (*Ulmus* 'Sapporo Autumn Gold') at the rear.

From the Dell Restaurant, head for **The Ring** and walk past it towards the **Store Yard**. Immediately facing you is a ring of ash trees of various species; these are grafted trees and the scions and the rootstocks are mostly sadly mismatched – some have huge stocks and thin stems and others the opposite.

From here, walk around the **Ranger's Lodge** into the **Lower Parkland**. Near the Lodge are two gnarled and ancient Sweet Chestnut trees (*Castanea sativa*) which, at possibly more than 200 years old, are probably the oldest trees in the Park. Between the Ranger's Lodge and the Serpentine is a group of tall Common Alder (*Alnus glutinosa*), while at its rear is a huge plane with an inexplicably bottle-shaped trunk. The trees in the remainder of the area consist mainly of a collection of oaks, both common and exotic species. The acorns of some of the native oak species have been converted into star-shaped galls.

The border facing the rear of the **Old Police House** has several unusual species. To the rear of the enclosure on the approach to **Rima** is a large Black Walnut. The small tree with very large, tough, evergreen leaves is a Chinese Evergreen Magnolia (*Magnolia delavayi*), whose gloriously scented, parchment-coloured flowers appear in late summer. Growing close to the fence is a

19

rare Chinese Varnish Tree (*Rhus potaninii*), a smallish tree with large, pinnate leaves, bearing seven to eleven leaflets and colouring vividly in autumn. There is also a large, tree-sized Bay Laurel (*Laurus nobilis*) a little further behind. Opposite Rima are two very rare specimens: look for a couple of trees with deeply furrowed bark and small, odd-looking crowns. These are Single-leaved False Acacias (*Robinia pseudoacacia* 'Unifoliola').

From here, continue steadily through the **Upper Parkland**, past flourishing limes and a healthy Indian Bean Tree (*Catalpa bignonioides*), towards the **North Carriage Drive** and the large mature Sycamore (*Acer pseudoplatanus*) in the corner. Sycamores are gawky as young trees, but attain dignity in late middle age. They also make an appalling nuisance of themselves as they distribute their unwanted seeds far and wide. To the left, near the **Victoria Gate**, is an unusually large Field Maple (*Acer campestre*) – good to see even in its winter silhouette.

Cross the North Carriage Drive and admire the boundary-planting, which contains several Weeping Beech, a fine Copper Beech and a Caucasian Wingnut opposite **Hyde Park Street**. Further on, by the **Albion Gate**, is a group of Pendent Silver Limes (*Tilia* 'Petiolaris') and, yet further on, a Tree Privet (*Ligustrum lucidum*). In front of the **Tyburn Convent**, and London's narrowest house, there are groups of various forms of wingnut and beech trees and, just before the **Cumberland Gate**, a young Golden Ash (*Fraxinus excelsior* 'Jaspidea'), which has glowing yellow leaves in autumn. On the Park side of the North Carriage Drive is a screen-planting of young American oaks, in this case Pin Oak (*Quercus palustris*).

From this point, head across the Park towards the **Reformers' Tree**, and again diagonally across the **Parade Ground**, to enjoy the lower part (the quietest section) of the Broad Walk; then cross to **Lovers' Walk**. Originally the Walnut Avenue, the trees were destroyed by a great frost in the eighteenth century and a replacement avenue of walnuts has been planted. Closer to the boundary can be seen some limes, including the American Lime (*Tilia americana*), and Indian Horse-chestnuts (*Aesculus indica*). The candles of this chestnut are whiter and appear a few weeks later than those of the common Horse-chestnut (*Aesculus hippocastanum*). On the approach to **Achilles Statue**, a low mound can be seen to the left, planted partly with silver-leaved species. This is the **1977 Silver Jubilee Planting**. In the centre of the shrubbery, partly concealed behind a birch tree, is one of the most reliable eucalypts for general planting, the Cider Gum (*Eucalyptus gunnii*). From here go straight towards Achilles, and note the sapling Cork Oak *(Quercus suber)* and Holm Oaks (*Quercus ilex*) planted at his back; these are intended to grow into an appropriate setting for the sculpture.

The walk ends at the Apsley Gate.

A Tree Walk in Hyde Park

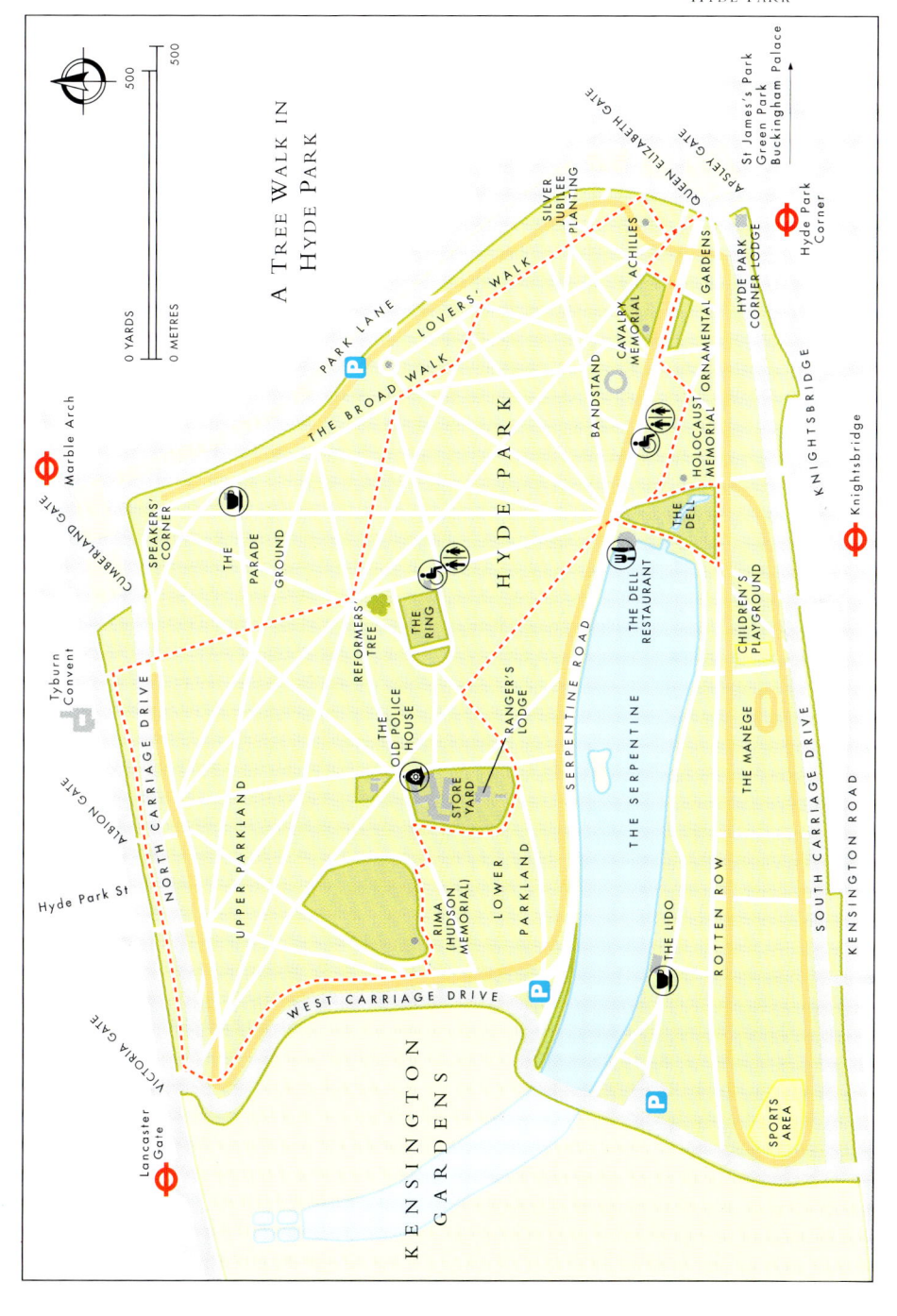

21

Kensington Gardens

Kensington Gardens

KENSINGTON PALACE has always had gardens. Their development had followed contemporary changes in taste fairly closely up to the time of the death of Queen Caroline (wife of King George II) in 1737, when their permutations came almost to a stop. The early, fussy, formal gardens, laid out by George London and Henry Wise in the seventeenth century, were swept away by Queens Anne and Caroline but the Round Pond and the *patte d'oie*, and its secondary system of cross-avenues, designed by Charles Bridgman, remain in place. William Kent, architect and landscape gardener, is reputed to have had something to do with the Gardens in the eighteenth century and Horace Walpole reports that: 'Kent, like other reformers, knew not how to stop at just limits... in Kensington Gardens he planted dead trees, to give a greater air of truth to the scene'. It is impossible to identify any of Kent's landscaping work in the Gardens today, apart from Queen Caroline's Temple, and, perhaps fortunately, Lancelot Brown was not involved in either their improvement or upkeep, because he had a habit of sweeping formality into oblivion.

Facing page:
Common Yew
Taxus baccata

Right: Weeping Beech
Fagus sylvatica 'Pendula'
South Flower Walk

There are tides of opinion in human affairs and this excerpt from a letter of Samuel Molyneux, dated 14 February 1717, shows how 'Taste' was on the move at that time:

> The gardens of Kensington are accounted a Master peice of Art in the new regular manner of greens and gravel gardening, for my part I must confess I have no opinion of this way at all, and tho I own that the gravel pit at Kensington is happily enough dispos'd, the antique Busts and Statues very well placed at the end of the little walks, tho the Mount made in Appearance on level ground by trees of different Heights, and the small compartment of high greens I observ'd with a Statue in the middle of it of about 40 or 50 yards Diameter with 8 walks centering there and as many seats let into the hedge between them tho I say all this be in it's way very agreeable yet in my opinion all this falls so low and short of the sublime unconfinedness of nature, and there is something so infinitely more exalting in the beautiful Scaravagie of noble grown Trees in a wild Wood that I cannot conceive how the world is so entirely fall'n into this way of Gardening.

The spelling and punctuation are those of Samuel Molyneux, as is the word 'Scaravagie'. This seems to be a version of a Chinese word, *sharawadgi*, made

Right:
Golden Indian Bean Trees
Catalpa bignonioides 'Aurea'
Serpentine Gallery

Facing page: Medlar
Mespilus germanica
Buck Hill

fashionable by the Scottish architect, Sir William Chambers. It is supposed to mean 'beauty of a highly sophisticated but irregular form'. Once thought to be bogus, it has been given respectability by modern Chinese scholars. Today there is a lot of 'scaravagic' tree-planting around, whereas the formality of the Gardens has survived by some kind of miracle.

Formal gardens went out of favour during the eighteenth century. In Kensington Gardens, which had survived with some of their historic landscaping still intact, the formal structural tree-plantings declined, not from deliberate destruction of the Bridgman plan but from what was, at the time, a general lack of sensitivity towards grand design. Time was to permit formality gradually to decay into more appropriate irregularity. There has now been another general change in sensibility and an interest has arisen in maintaining historic structures and landscapes in a proper manner. The task of restoring the Gardens to something of their past glories has been started and is showing promising progress.

When the work of fully restoring the original lines of the avenues began about ten years ago, it was found that, over the centuries, and for a variety of reasons, some lines had become up to several metres out of alignment and the axial lines no longer converged at the original focal point. Gaps had been replanted with inappropriate species, drainage systems (all of which are notorious for their limited life) had become blocked, and the rising water table was drowning the roots of some of the trees. A document describing the appropriate actions necessary to put right these, and other, matters, was prepared in 1988/9 and work began, appropriately, with the replanting of the Great Bow.

The re-instatement is now approaching completion and visitors will discover many young trees – as single specimens, in lines and in regiments.

In the great storm of 1987, the Jubilee Walk was severely damaged and had to be completely replanted. The replacement trees, Silver Limes (*Tilia tomentosa*), were donated by the Citizens of Berlin as a gesture of friendship. These have now been in the ground for several years and are making good progress.

A different problem arose with the Broad Walk. This had become geriatric and, after a good deal of controversy, was completely replanted with inner lines of Norway Maples (*Acer platanoides*) and outer lines of Broad-leaved Limes (*Tilia platyphyllos*) in 1953.

The little rococo gardens in the spaces between the avenues, conventionally called the Quarters, were swept away centuries ago. They could never have been anything but a muddle and a problem for the Head Gardener but were, for a long time, given individual treatment. They were allowed to fall into decay in the same way as the avenues but are now also being re-instated. The collection of chestnut species in the Chestnut Quarter is being brought up to strength and, in the adjoining quarter, a collection of Sweet Gums (*Liquidambar styraciflua*) is presently being assembled – and may eventually become a National Collection. The autumn colour of Sweet Gums can be rich and fiery and, in good years, may be spectacular.

What Samuel Molyneux called 'the Mount made in Appearance on level ground' (see p.24), and Joseph Addison, owner of *The Spectator*, referred to as a 'Seeming Mount', as well as the adjoining sunken garden, known at the time as the Pit, were admired extremely before their demise. The Pit cannot be redug at the present time but restoration of the Seeming Mount is a feasibility; it could again be a popular feature of the gardens and so the possibility of financing the work and the details of its planting are being explored. Records show that, in 1715, Henry Wise, the King's Gardener, was paid £30 for making a moving scaffold for 'cutting the trees' in the Gardens and this may well have been used to high-prune the trees on the Seeming Mount.

Kensington Gardens is a rich site for nature conservation: the large area of undisturbed ground around Long Water, and between Long Water and the boundary of Buck Hill, is home to many native trees and herbaceous plants, with access for thirsty creatures to fairly clean water, and good living conditions for birds and animals. While nest-boxes have been put up in most of the Royal Parks, in these Gardens partially hollow trees (closely monitored for safety reasons) have been retained as well. These not only provide roosting sites for larger birds but also serve as roosts for the declining bat populations. A recent initiative has been the provision of a few heron platforms in safe trees.

A Tree Walk

Enter the Park by the **King's Arms Gate**. On the sloping ground to the left is a small group of young limes, planted with the intention of rendering the large building alongside a little less prominent. Further along the western boundary of the Park are some old, but not ancient, Black Mulberry trees (*Morus nigra*), planted against the wall. Although denoted **Palace Avenue** on the map, this is usually known as the Mulberry Walk. Growing close to the southern fence bounding **Kensington Road** are a number of Indian Horse-chestnuts (*Aesculus indica*), a more elegant species than the European tree, with more vertical, more slender, whiter flowers and less coarse leaves. The **Dial Walk** has been replanted with Tulip Trees (*Liriodendron tulipifera*), which should mature into a fine approach to **Kensington Palace**; broadly columnar, they have butter-yellow autumn colouring. In the open lawn, close to the **Studio Gate** path, is a Paulownia (*Paulownia tomentosa*), a spectacular tree when in bloom, with startling clusters of foxglove-like, lilac flowers, and also conspicuous in winter, when its twigs are beaded with embryo flower buds.

Make your way around the Palace and, after a detour to inspect the **Sunken Garden**, look at the Bay Laurels (*Laurus nobilis*) and the fine, clipped hollies between the Palace and the **Orangery**. Both sexes of the holly are represented here, the female plants being conspicuous in winter, with their show of scarlet berries. The hollies have more 'presence' than the pyramidal thorns alongside. Between the Orangery garden and the **Broad Walk** there is, firstly, another Paulownia and, secondly, nearer the Orangery, a Black Walnut (*Juglans nigra*).

Tulip Trees
Liriodendron tulipifera
Dial Walk

Date Plum
Diospyros lotus
South Flower Walk

Walk past the Orangery, inspecting the two very rare trees behind it: the furthest from the path is a Date Plum (*Diospyros lotus*) and the nearest is a Euodia (*Tetradium daniellii*), whose ash-like leaves are conspicuous in early autumn when the branches bear heads of yellowish white flowers. Continue along the Broad Walk to the **Bayswater Road**. Near the **Black Lion Gate** is a Weeping Ash Tree (*Fraxinus excelsior* 'Pendula'), whose branches meet the ground, making the tree a favourite place for children to play 'house'. Beside it is a Schwedler's Maple (*Acer platanoides* 'Schwedleri'), with dark leaves and sometimes good autumn colour; this species is planted here and throughout most of the Royal Parks. There are a few promising young specimens of various American oaks, notable for their sharply pointed leaf lobes, but gapping-up to restore the consistency of the planting is nowadays done with limes. It is worth taking a detour opposite the Park Court Hotel and to walk about 300 paces into the open Park to see a magnificent specimen of Red Oak (*Quercus rubra*).

Turn eastwards along the **North Walk**. The tree screen here was once a decaying planting but is now being gradually re-instated. Its continuation, known as the **North Flower Walk**, is being variegated with more interesting species in order to relieve some of the pressure on the **South Flower Walk**. While walking along the northern boundary, keep an eye open to the right so as not to miss the grand vistas down the avenues: the **Inverness Walk**, with its lines of Common Lime (*Tilia* x *europaea*), and the **Lancaster Walk**, planted with London Plane (*Platanus* x *acerifolia*); the tree lines are particularly fine.

The North Flower Walk is bounded by a low 'hairpin' railing. Enter the gate by the Forsythia Shrubbery. This is marked by a plaque indicating that the planting was in commemoration of the great arborist and horticulturist, William Forsyth, who described himself as 'Gardener to His Majesty at Kensington and St James', and also gave his name to the shrubs growing on the site. Walk on, inside the enclosed fence, until you come to a mature Austrian Pine (*Pinus nigra* subsp. *nigra*) and a Keaki (*Zelkova serrata*) on the other side of the path, followed by a young Red Oak and a healthy, dark-foliaged Holm Oak (*Quercus ilex*). The next evergreen tree, an unusual species to find in a park, is a fairly large specimen of Common Yew (*Taxus baccata*). A little further along is a Ginkgo (*Ginkgo biloba*), a deciduous tree with fan-shaped leaves and a history dating back over two million years; by its upright habit it appears to be male. Carry on to inspect another Weeping Beech and an unusually tall Bay Laurel nearby. There is a Medlar (*Mespilus germanica*) near the junction of the paths leading to the **Marlborough Gate**. A tree too seldom planted, the Medlar's boughs have an attractive wriggly habit, like a tree drawn to illustrate a fairy-tale, and, in early spring, bear delicately poised, large, cup-shaped white flowers.

A Dawn Redwood (*Metasequoia glyptostroboides*), showing signs of outgrowing its neighbours, stands in front of the public conveniences and,

Black Mulberry
Morus nigra
Mulberry Walk

Norway Maple
Acer platanoides
Broad Walk

beyond **Queen Anne's Alcove**, there is a Montpelier Maple (*Acer monspessulanum*). Near **Buck Hill**, on the open lawn, is another, even bigger Medlar. From here, after admiring the Victorian extravanganza of the **Italian Garden**, turn towards the **Peter Pan enclosure**. To the right, on the open lawn, is a small group of recently planted Indian Bean Trees (*Catalpa bignonioides*). On the edge of **Long Water** are the species one would expect to find: a tall, blunt-topped Swamp Cypress (*Taxodium distichum*), tall, raggedly picturesque Common Alders (*Alnus glutinosa*) and, much further along the path, a large Black Poplar (*Populus nigra*), Golden Weeping Willow (*Salix* 'Chrysocoma'), Babylon Willow (*Salix* x *sepulcralis*) and White Willow (*Salix alba*).

The Peter Pan enclosure contains a couple of interesting exotic species. Leaning over the fence at the corner on the approach is a rare Variegated Tree Privet (*Ligustrum lucidum* 'Excelsum Superbum'), whose leaves are ornamented with white and cream; it flowers exuberantly in late summer for a long period. The other tree of interest stands behind and to the left of Peter Pan's statue; this is the Judas Tree (*Cercis siliquastrum*), with rounded leaves and masses of quaint, stalkless, carmine flowers in late spring. Just past the enclosure, to the right, is a splendid old Sweet Chestnut (*Castanea sativa*), possibly a survivor of the 1734 planting.

From the end of Long Water, head for the **Serpentine Gallery**, where, on the east lawn, are two fine, large Golden Indian Bean Trees (*Catalpa bignonioides* 'Aurea'), whose leaves open bronze but become bright yellow in autumn. Nearby is a group of Caucasian Limes (*Tilia* x *euchlora*). From here, walk in a straight line southwards to the Flower Walk, following the path around the rear of the public conveniences to find a Deciduous Camellia (*Stuartia pseudocamellia*) by the **East Snake Walk**. With advancing age, this tree develops a smooth bark, in shades of cinnamon, cerise, pink or even purple; this fairly young specimen has yet to achieve its full development. Now join the path which runs parallel with the Flower Walk. Nearby are a number of attractive trees which have overgrown the walkway to partially enclose it. There is a fine Weeping Beech, a Holm Oak and a partially decapitated Lebanon Cedar (*Cedrus libani*). Before heading for the Flower Walk, carry on towards the main road and over the **Albert Memorial Road**. Near the boundary and the corner of the Albert Memorial is a mound of refined evergreen foliage; this Syrian Maple (*Acer syriacum*) may not be a flamboyant plant but it is probably the rarest species in the Park.

From here, walk straight into the Flower Walk. Behind the first seat recess, opposite a cedar, is a hornbeam – and a good view of the **Prince Albert Memorial**. Further along is another cedar and, midway on this stretch, a large Californian Laurel, or Headache Tree (*Umbellularia californica*) – a good sniff at the leaves will instantly bring on the most appalling headache, so please be warned; the agony is fierce. To the left, about two-thirds of the way along the Walk, are a Holm Oak, a Ginkgo and, with deeply fissured, corky bark, a Cork Oak (*Quercus suber*). Near the gate into **Snob's Crossing** another Holm Oak stands to the right and a Deodar (*Cedrus deodara*) to the left.

Cross Snob's Crossing and enter the western end of the Flower Walk. Immediately inside, to the right of the gate, is a yew which has been invaded

Weeping Beech
Fagus sylvatica 'Pendula'
North Flower Walk

by a wisteria vine in an interesting fashion. About 20 paces further on is the Weeping Beech in which Peter Pan is reputed to have climbed and about halfway into the Walk, to the left, is another Date Plum. Date Plums are either male or female and since this one, which bears (and sheds to the pigeons) showers of raspberry-sized, yellow and pink fruits, is female, the other Date Plum near the Orangery is probably male. Just beyond the two Chusan Palms (*Trachycarpus fortunei*) to the right is a Weeping Ash (*Fraxinus excelsior* 'Pendula'). The two gems of this walk are the mature, fruiting Date Plum and the single, shapely evergreen standard tree set in the lawn in front of the left-hand flower border near to the exit – a specimen of *Osmanthus heterophyllus*, which bears clusters of fragrant white flowers in the axils of the leaves in early autumn. It is a rare and precious thing.

From the gate, turn towards the junction of the **West Snake Walk** and South Carriage Drive. At this point you should have no problem in identifying the contoured tree with shedding, pale brown bark as a Persian Ironwood (*Parrottia persica*). Here, from the lawn, is a good place to view the magnificent planes which were formally planted to frame the Albert Memorial.

The walk ends at the exit from the Flower Walk.

Osmanthus heterophyllus
South Flower Walk

Kensington Gardens

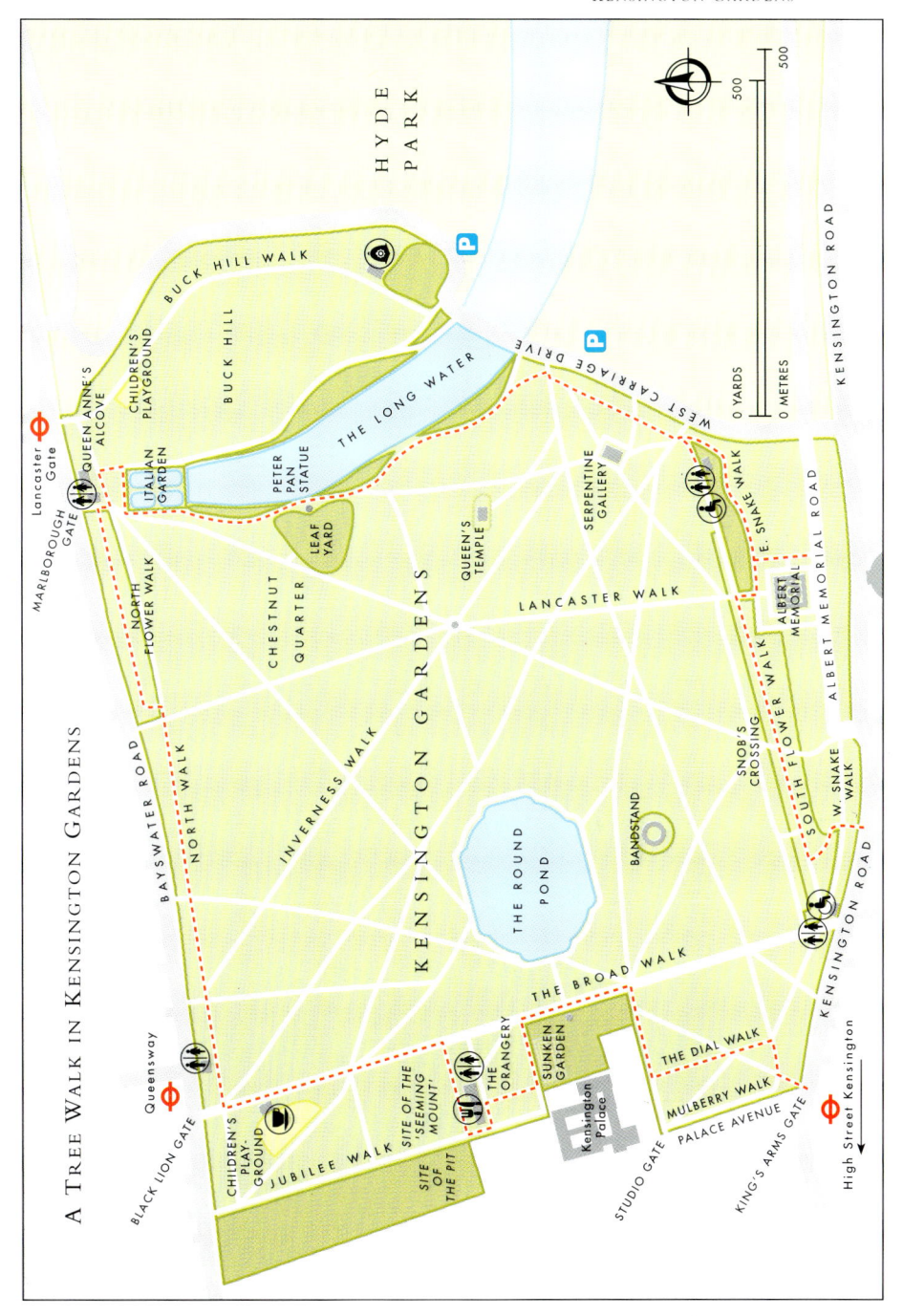

St James's Park and Green Park

St James's Park and Green Park

These two Parks have suffered more design permutations than any other of the Royal Parks. Originally a stretch of marshy rough pasture for pigs, with the sluggish Tyburn Stream flowing through the middle, and, in its upper reaches, a burial ground for the St James the Less Leper Hospital, it was acquired by the sport-loving King Henry VIII in 1531. The Parks have, at one time or another, contained: a zoo, a bird collection (St James's Park still does), a silk farm, a *paille-maille* (a type of croquet) court or courts, a tilt-yard, a bowling alley, a physic garden, a bird-watching hide, a duck decoy, a reservoir, a dairy herd and an ice-house. They have also been a venue for celebratory firework displays incorporating fantastic temporary architecture; these, because of the very nature of fireworks, sometimes became gigantic, spectacular bonfires! Dig where you will and rubble, foundations or mingled earths appear. This is not an obviously promising site for tree growth but, in spite of the soil and the polluted atmosphere, there are some splendid trees in both Parks.

From time to time, the tree populations suffered severely from the activities described and, during the days of the Commonwealth (1649–1660), many were cut down to use as fuel. Today, however, the Parks are London's jewel, an unexpected paradise, swarming with pilgrims and tourists from dawn to dusk.

Experiments in novel tree species were attempted as long ago as the seventeenth century, when Edmund Waller, the poet and politician, lauded the

Facing page:
London Plane trees
Platanus x *acerifolia*
Broad Walk, Green Park

Right: Tibetan Cherry
Prunus serrula
St James's Park

35

Left: Indian Bean Tree
Catalpa bignonioides
St James's Park

Below: Black Mulberry
Morus nigra
St James's Park

plantings carried out by André Mollet around the new canal (today transmogrified into St James's Park Lake):
   …young trees upon the banks
   Of the new stream appear in even ranks.
   The voice of Orpheus or Amphion's hand
   In better order could not make them stand.
   ('On St James's Park As Lately Improved by His Majesty', 1661)
Unfortunately, all the trees planted were False Acacias (*Robinia pseudoacacia*) and neither of the above immortals were able to make them stand; the breath of Boreas would keep blasting off the side branches and so the trees were all removed and replaced. A few have crept back into the plantings but the species remains a brittle one.
   Much of the structural tree-planting of the avenues and formal groups was, for a long period, of Common Elm (*Ulmus procera*) but all of them vanished a

very long time ago. Elm disease is not a new thing, as this extract from a piece entitled 'Discovery of the Secret Destroyers of the Trees in St James's Park', by 'Dendrophilus' (1823) shows:

> In spring we see the leaves sprout forth from the venerable branches in all the luxuriance of vegetation, when of a sudden they are blasted as if by lightening, the bark falls from the stem, and long ere winter the finest tree in the park is only fit for firewood... Now every elm is to extent infected, and every week we may observe that a tree has perished.

This is an excellent description of the progress of elm disease. It can be no coincidence that most of the oldest and largest plane trees in the central London parks date from not long after 1823, when presumably the decision was taken to replace the elms with more disease-resistant species. Those trees in St James's Park itself, involved in John Nash's re-landscaping of about 1827, are unlikely to be older than this date.

St James's Park has the higher density of exotic species, and the more romantic landscape – interspersed with patches of brilliant horticulture. Green Park – a quiet green sanctuary, away from the noise and fumes of the surrounding city – has its devoted adherents. Both Parks are framed by formal rows of plane trees; the largest grow some distance away from the pollution of petrol fumes and where winter salt-dressings, washed from the roads, cannot penetrate the root areas. Those bordering The Mall are given periodic 'haircuts' in order to leave the highway free for traffic. Planes dominate both Parks: in St James's Park they provide an open canopy that protects the ground plantings and, in Green Park, a dappled shade on the almost unbroken green stretches of lawn.

London Plane
Platanus x *acerifolia*
Green Park

A Tree Walk

Enter **St James's Park** from **Queen Anne's Gate** and walk towards the bridge. To the right is the **Leaf Yard** and to the left is a low artificial mound. There are several Black Mulberry trees (*Morus nigra*) of various ages on the mound and one near the fence, planted to remind the passer-by of King James I's failed silk industry. Silkworms will eat Black Mulberry leaves, as well as fresh oak leaves, but neither of these diets produces good silk. For the best silk, silkworms should be fed on leaves of the White Mulberry (*Morus alba*), which is more difficult to grow. Perhaps poor King James I was not well advised! In the left corner of the lawn, near the bridge, is a Weeping Beech (*Fagus sylvatica* 'Pendula').

Nearer **Birdcage Walk** is a Walnut (*Juglans regia*) and, a little further on, a Tree of Heaven (*Ailanthus altissima*) has been planted to provide a contrast in foliage. On the edge of **St James's Park Lake** is an excellent Swamp Cypress (*Taxodium distichum*) beyond which, standing on the lower side of the path, is

Facing page and above:
Judas Tree
Cercis siliquastrum
St James's Park

Right: Tree of Heaven
Ailanthus glandulosa
St James's Park

a mature Winter-flowering Cherry, probably *Prunus subhirtella* 'Autumnalis Rosea'. From here, walk on to find, near the playground, yet another Swamp Cypress and, on the bank at the head of the lake, a large Copper Beech (*Fagus sylvatica* 'Purpurea').

Change direction at this point and head straight for the **Canada Gate** entrance into **Green Park**, where you can view the splendid avenue of majestic plane trees. There are some exotic maples and two raggedly picturesque poplars east of the **Constance Fountain** but few casually planted trees of any great botanical or historical interest – except one. This can be found near the boundary with **Piccadilly**, just off the northeastern side of the low mound which is all that remains of the **Snow House** (presumably an ice-house). The tree is opposite the end of **Down Street** and appears to be the cultivar of the plane known as *Platanus* x *acerifolia* 'Augustine Henry' (see also p.12). It is already a very large tree but is still actively growing. The leaves are twice as large as those of the common London Plane, some being 38cm (15in) wide, and of a better green, and the bark has a smaller pattern of flakes. It seems to be even better adjusted to the appalling pollution of central London than the majority of planes.

Return to **St James's Park** and resume the walk around the lake. Towards the Buckingham Palace end of the lake another Swamp Cypress stands at the

Scarlet Oak
Quercus coccinea
St James's Park

water's edge and, nearer to **The Mall**, there are two Dawyck Beeches (*Fagus sylvatica* 'Dawyck'), still pencil-thin. These will become more thickly columnar as they age. A little past the junction of two paths, in the middle of the green, is a Scarlet Oak (*Quercus coccinea*), with a planted bed containing some fairly unusual species near the path. Here you will find a Californian Laurel, or Headache Tree (*Umbellularia californica*), a Golden False Acacia (*Robinia pseudoacacia* 'Frisia'), a eucalyptus and some attractive shrub underplanting. The bed tucked into the angle of the bridge path contains a fine specimen of Holm Oak (*Quercus ilex*). Further up the slope, to the left of the path, is what appears to be another 'Augustine Henry' cultivar of the London Plane while, by the water, you can see one of the large Figs (*Ficus carica*) that grow so rumbustiously in the Park. Continue walking along the lakeside towards the **Cake House**.

In the middle of the lawn, near the Cake House, is a Red Oak (*Quercus rubra*), while further on there are two more young Dawyck Beeches. On a breezy day, the silvery-grey undersides of the leaves of the Silver Lime (*Tilia tomentosa*) that stands near the exit from the Park to **Admiralty Arch** provide a scintillating display. To the right of the path, in the middle of the fenced lawn, is a Silver Pendent Lime (*Tilia* 'Petiolaris'), a relation of the Silver Lime; this is a much older tree, with a conspicuous graft union.

Continue around the perimeter path to the **Store Yard** and then press on beyond it. The screen-planting around the Store Yard includes a Judas Tree (*Cercis siliquastrum*) and a thriving mixed border of shrubs, herbaceous plants and seasonal bedding. Near **Duck Island**, where the path and grass verges start to narrow, there is a splendid Medlar (*Mespilus germanica*) and, just where the paths diverge, a Tibetan Cherry (*Prunus serrula*), with shining brown-boot-polished bark. An Indian Horse-Chestnut (*Aesculus indica*) stands nearby and there are a number of vigorously growing fig trees near the water.

Between this point and **Storey's Gate** are two Euodias. Walk on around the perimeter-road path to find, at the rear of the Leaf Yard, two specimens of Golden Rain Tree, or Pride of India (*Koelreuteria paniculata*). This beautiful species should be planted more widely: a three-season tree, its unfolding leaves are bright pink in spring, profuse panicles of yellow flowers appear in late summer and, in a good year (and a good tree), the foliage becomes orange and the inflated seed capsules turn red in autumn.

Duck Island is planted with a mixture of species commonly found in such situations: Lombardy Poplars (*Populus nigra* 'Italica'), Grey Poplar (*Populus canescens*), White Willow (*Salix alba*), Golden Weeping Willow (*Salix* 'Chrysocoma') and Babylon Willow (*Salix* x *sepulcralis*). These also occur here and there around the margins of the lake.

The walk ends at Queen Anne's Gate.

Right: Silver Lime
Tilia tomentosa
St James's Park

Below: Paper-bark Maple
Acer griseum
St James's Park

St James's Park and Green Park

Green Park

PICCADILLY

GREEN PARK

DOWN ST

Hyde Park
Kensington Gardens

Hyde Park

CONSTANCE FOUNTAIN

SITE OF SNOW HOUSE

THE BROAD WALK

CANADA GATE

CONSTITUTION HILL

Buckingham Palace

Victoria

0 YARDS 300
0 METRES 300

42

A Tree Walk in St James's Park and Green Park

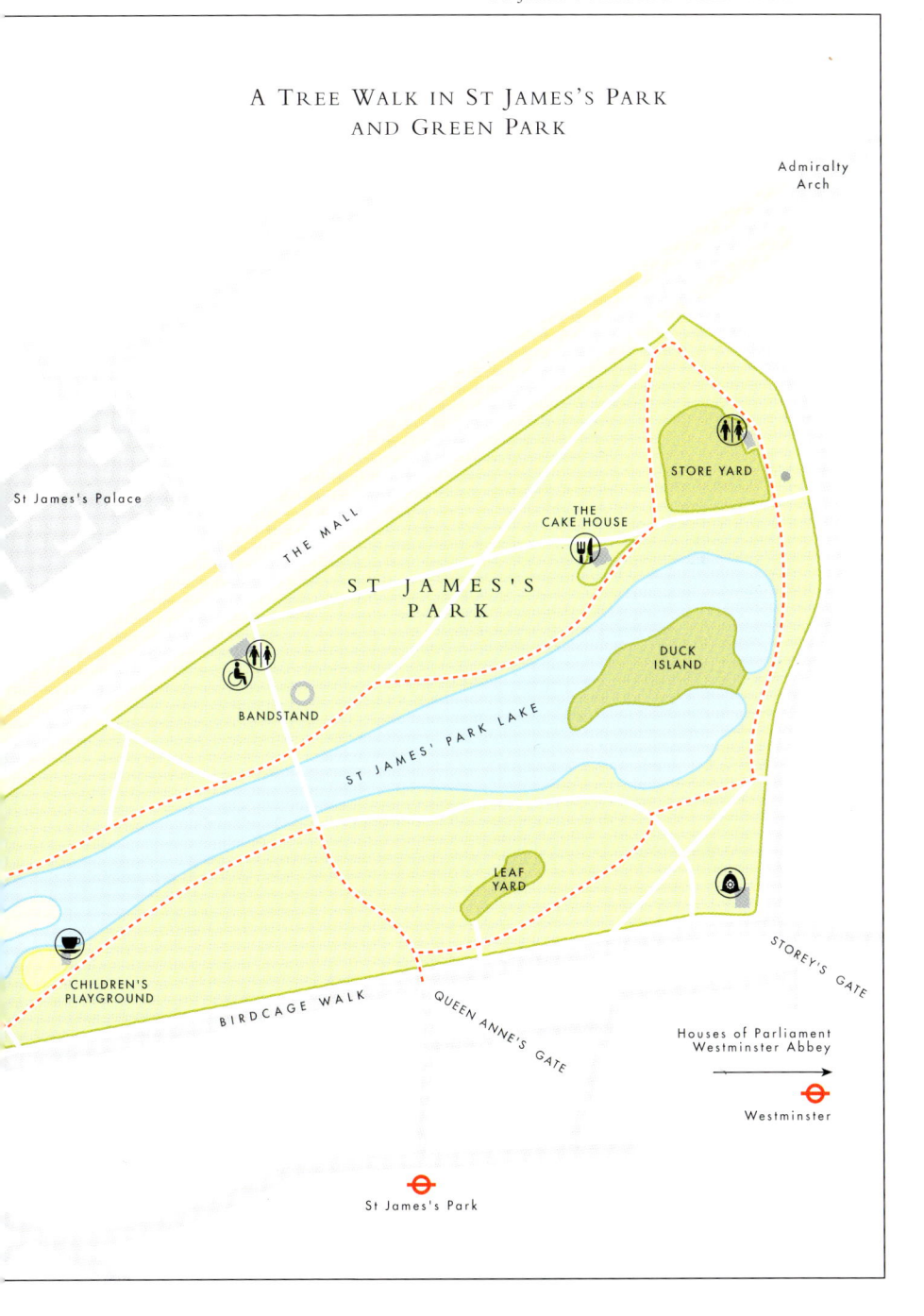

43

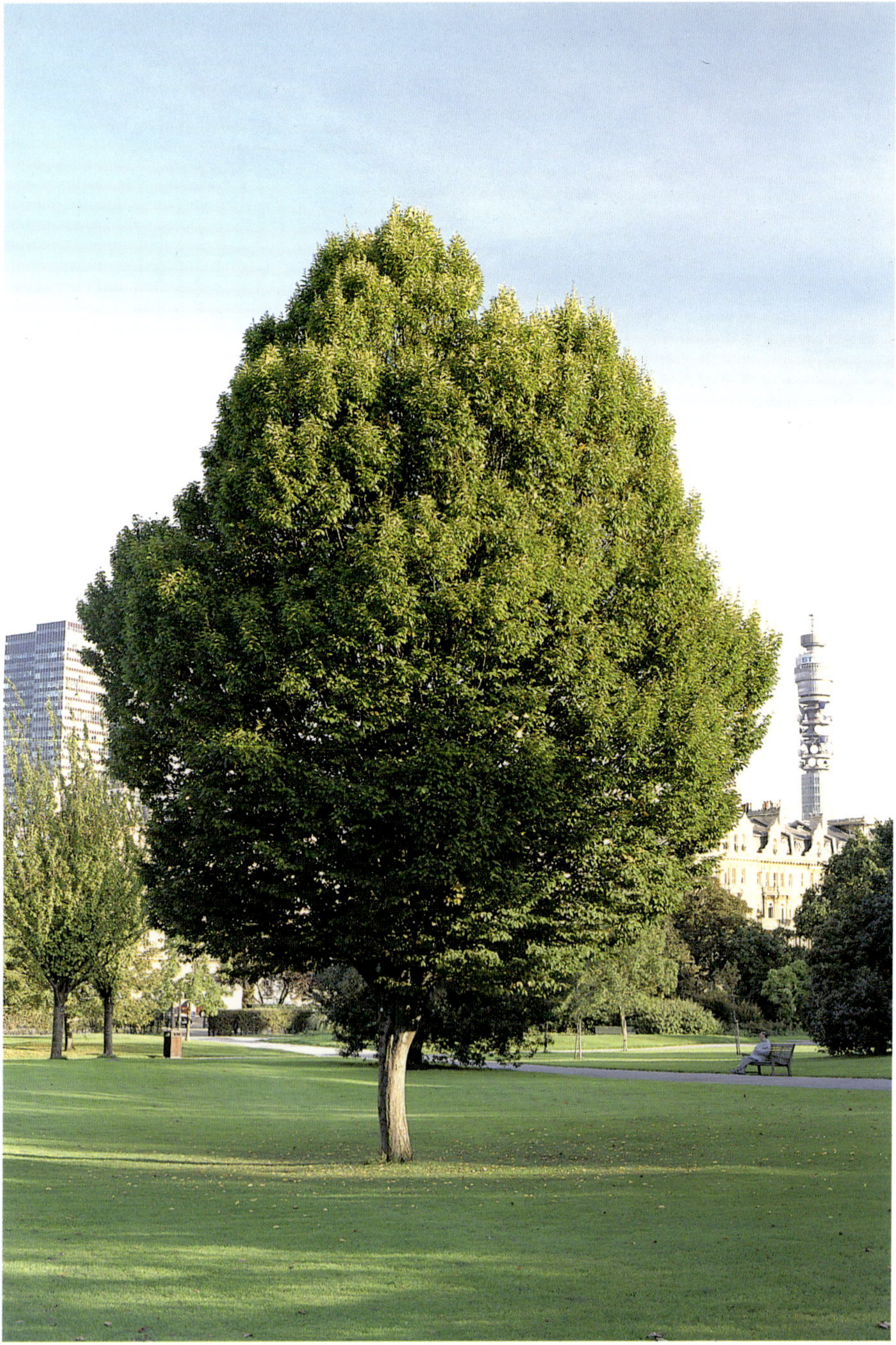

REGENT'S PARK

THIS IS yet another of the Royal Parks which were acquired by Henry VIII – in this case in 1539 – and it functioned as a Royal Hunting Park for about a century. In its early years it was known as Marybone Park and, later, as Marylebone Park. The deer disappeared very early on and the area was formally disparked in 1668. This, however, did nothing for its tree population. The land was leased for farming almost until the time of Nash's landscaping in the early nineteenth century. A survey made at some time during the Commonwealth (1649–1660), showed that the timber in the Park was worth the equivalent of £1,774.40 in present-day money. This timber was reserved for the Navy and was felled and removed from the site in 1649; a general clearance of all sorts of trees by local people followed, both authorised and unauthorised. Thus, Marylebone Park remained almost treeless until Nash, in about 1811, started his great scheme of civic planning, an important part of which was the creation of a new park for the Prince Regent.

The site, over an enormous depth of London clay, was very wet. The Tyburn Stream wandered over the fields, which had been used for a variety of industrial purposes: one field had been partly ruined by gravel extraction (and was probably incorporated, by Nash, into the Lake) and another area had been used for the manufacture of saltpetre, which involved massive quantities of 'manure'.

Facing page:
Fastigiate Hornbeam
Carpinus betulus 'Fastigiata'
English Garden

Right: Lime trees
Tilia cordata 'Greenspire'
Avenue Gardens

45

The site geology, and its usage and abusage over the last 300 years, including the deposition of enormous depths of rubble as a result of the clearance of bomb damage in the course of the Second World War, have constituted problems which have bedevilled all Park Managers and their assistants ever since Marylebone Park became Regent's Park and a place of public resort.

Soil drainage and amelioration have continued unceasingly for the last 150 years and some fine trees have now developed, although none can be described as ancient. John Nash redesigned St James's Park in a fairly short time, leaving a fully harmonious landscape. Regent's Park is different: a great deal of the design was intended to be townscape but Nash's ambitions were largely frustrated by a downturn in the international economy. As a consequence, the only fully completed Nash landscape is in the vicinity of the Lake. A good deal of the rest of the Park has 'just growed', although, for the last century or so, the Park has had some remarkable Superintendents, who have raised its planting and management to a very high level.

The oldest exotic and more interesting species are to be found around the Lake, in Queen Mary's Gardens, and in and around the William Nesfield Avenue Gardens. This is also where the most sumptuous flower displays occur.

TREE WALKS

The tree walks are in two parts but energetic walkers can traverse them both in half a day, as the second walk starts where the first leaves off.

QUEEN MARY'S GARDENS

Enter the Park from the **Clarence Gate** and walk around the path to the bridge. Near the edge of the **Lake** are weeping willows, White Willow (*Salix alba*) and Silver Willow (*Salix alba* f. *argentea*), poplars, including Lombardy Poplar (*Populus nigra* 'Italica') and alders. Cross the bridge and turn left; immediately inside the fence of Regent's College is a large Crack Willow (*Salix fragilis*) and a Common Hornbeam (*Carpinus betulus*). There are masses of all these trees on the island and, on the bank close to the bandstand, there is a group of hybrid Balsam Poplars (*Populus balsamifera* x *trichocarpa*, probably Clone 32).

Cross the **Inner Circle** into **Queen Mary's Gardens** by the path alongside the **Rose Garden Café**; here there is a huge plane, immediately in front of which are several Lombardy Poplars, partially surrounding the **Begonia Garden**. Beneath the poplars are some young Honey Locusts (*Gleditsia triacanthos* 'Sunburst'), a cultivar which has golden young shoots throughout summer. Between the poplars and another huge plane is a very rare tree indeed, a Euodia (*Tetradium daniellii*), whose large reddish orange and lemon-yellow blossoms appear in late summer and remain on the tree for a long time. On the lawn as you walk around to the front of the Begonia Garden, there is

Facing page: Ash trees
Fraxinus spp.

Right: Holm Oak
Quercus ilex
Queen Mary's Gardens

Right: Paulownia
Paulownia tomentosa
English Garden

Left: Weeping Ash
Fraxinus excelsior 'Pendula'

one of a number of Atlas Cedars (*Cedrus atlantica*) and, to the left of the viewing gap in the hedge of the Begonia Garden, a small group of Cider Gums (*Eucalyptus gunnii*). On the planted mound to the other side of the lawn stands a mature Turkey Oak (*Quercus cerris*), a False Acacia (*Robinia pseudoacacia*), an old Silver Maple (*Acer saccharinum*) and a very gnarled Indian Bean Tree (*Catalpa bignonioides*) and, in the lawn itself, a Deodar (*Cedrus deodara*). Beside the exit to the Inner Circle, to the right of the path, is an old Black Mulberry (*Morus nigra*). Mulberries develop masses of heavy twiggy branches and bark with deep fissures, burs and rugosities; few escape damage of some kind and this one has had to be severely pruned.

Near the **Triton Fountain**, behind the yew hedge to the right, is a row of bottom-worked Yoshino Cherries (*Prunus* x *yedoensis*), which are stunning in early spring when they are crowded with dainty pink blossoms. The conifers backing on to the small pool and bog-garden are Dawn Redwoods (*Metasequoia glyptostroboides*) while, on the other side of the pool, there is a Corkscrew Willow (*Salix matsudana* 'Tortuosa') and a Ginkgo (*Ginkgo biloba*). Nearby, with beautiful, partially flaking grey bark with orange fissures, is a Caucasian Elm (*Zelkova carpinifolia*). There is another Honey Locust at the back of the herbaceous border and a young Dove, or Handkerchief, Tree (*Davidia involucrata*) by the seating area, with a Judas Tree (*Cercis siliquastrum*) nearby. Closer to the exit and the old Service Yard is a flourishing Kentucky Coffee Tree (*Gymnocladus dioicus*), with bipinnate leaves, and, nearer to the wall, a very attractive white-stemmed birch. While walking around to the exit – note what is, for this site, a very large, old Copper Beech (*Fagus sylvatica* 'Purpurea') and what appears to be a Cappadocian Maple (*Acer cappadocicum*). Alongside the statue of the **Mighty Hunter** is a variety of American Oaks.

Head for the path leading past the public conveniences where there is a large Tree of Heaven (*Ailanthus altissima*) close to a large Red Oak (*Quercus rubra*) and a Grey Poplar (*Populus canescens*). To the left of the path is a row of very impressive Bay Laurels (*Laurus nobilis*), which, shortly before 1982, were measured by Alan Mitchell and found to be 10m (33ft) tall. At the end of the row is a Caucasian Elm, also measured by Alan Mitchell and found to be 29m (95ft) tall.

This short walk ends at the nearby gate into the Inner Circle.

ENGLISH GARDEN TO QUEEN MARY'S GARDENS

From **Queen Mary's Gardens**, cross the **Inner Circle** to **Chester Road** and enter the **Avenue Gardens**. These have recently been restored to their former glory but it will be some time before the recently planted trees are mature. The trees of the inner avenue are limes and those of the outer avenue are Tulip Trees (*Liriodendron tulipifera*).

Walk around the back of the low mounds, which were planted by Markham Nesfield in the middle of the nineteenth century; some of the trees date from the original planting. In the lawn between the mounds and the border backing on to Chester Road is a middle-aged Paulownia (*Paulownia tomentosa*), a beautiful tree with spectacular lilac flowers on leafless branches in mid- to late spring but, like many plants with hollow or pappy-centred shoots, liable to sudden death. This specimen looks reassuringly healthy.

In the border and the planted mound are fine large hollies, which look vigorous enough to attract tree-measurers in the future! There is a tall Holm Oak (*Quercus ilex*), some large yews and, at the northern end of the mound, a splendid Tree Privet (*Ligustrum lucidum*), with shining leaves, 15cm (6in) long and correspondingly wide. Behind the weeping willow in the lawn stands a young Blue Cedar (*Cedrus atlantica* f. *glauca*). In the centre of the mound is a large deciduous tree, a queer specimen of the Tree of Heaven, whose pinnate leaves have a hugely enlarged leaflet at the end, with exaggerated lobes and glands. A younger tree of the same species stands to the south and is of the

Common Fig
Ficus carica

normal type. Further along are large Turkey Oaks (*Quercus cerris*), tall Lombardy Poplars (*Populus nigra* 'Italica') and an unusually large Babylon Willow (*Salix x sepulcralis*). Near the southern end of the Garden, between the mounds, is a Pear Tree (*Pyrus* spp.) and a Judas Tree (*Cercis siliquastrum*) and, near the gate in the corner, is a fine Schwedler's Maple (*Acer platanoides* 'Schwedleri').

Walk towards the main gate past a nearby ash tree which appears to be a Narrow-leaved Ash (*Fraxinus angustifolia*), a very graceful species. Retrace your steps along the Avenue Gardens, pass through the gate and enter the main length of the **Broad Walk**. This was originally planted with some homogeneity but this has been lost for a very long time and the plantings as far as the corner of the **Zoological Gardens** are a mixture of species. A start has been made in re-aligning and reforming the Broad Walk's original impressive regularity, using Common Oak (*Quercus robur*).

Go back to the Chester Road and cross into the sports area. Just inside the gate is one of the best Common Oaks (*Quercus robur*) in the Park and a Horse-chestnut (*Aesculus hippocastanum*). To the right is a collection of Black Walnuts (*Juglans nigra*) – which occasionally defy the squirrels and carry a crop of nuts – and alongside these are three shapely Norway Maples (*Acer platanoides*).

Return through the gate and the side entrance of Queen Mary's Gardens and go through to the Rose Garden. The three rocketing conifers near the boundary hedge are Douglas Firs (*Pseudotsuga menziesii*) and, on the margins of the pond, are Swamp Cypresses (*Taxodium distichum*), which are developing 'knees' (pneumatophores) that protrude through the surface of the water.

Near the boundary of the Gardens, opposite the first Swamp Cypress, there is a fine young Scarlet Oak (*Quercus coccinea*) and the weeping willow on the far bank of the pond has silvery leaves more like a White Willow than a Babylon Willow. On the near bank, the small tree with pendulous branches and willow-like, silvery leaves is a Willow-leaved Pear (*Pyrus salicifolia*), a long-lived species. The large Golden Weeping Willow (*Salix* 'Chrysocoma') towards the entrance gate is developing a wonderfully fissured bark and there is a tall Weeping Ash (*Fraxinus excelsior* 'Pendula') near the open lawn.

Walk back to the splendid entrance gates, donated by the artist Sigismund Goetze, admiring the flower displays on the way and end the walk at this point.

Babylon Willow
Salix x *sepulcralis*
Queen Mary's Gardens

REGENT'S PARK

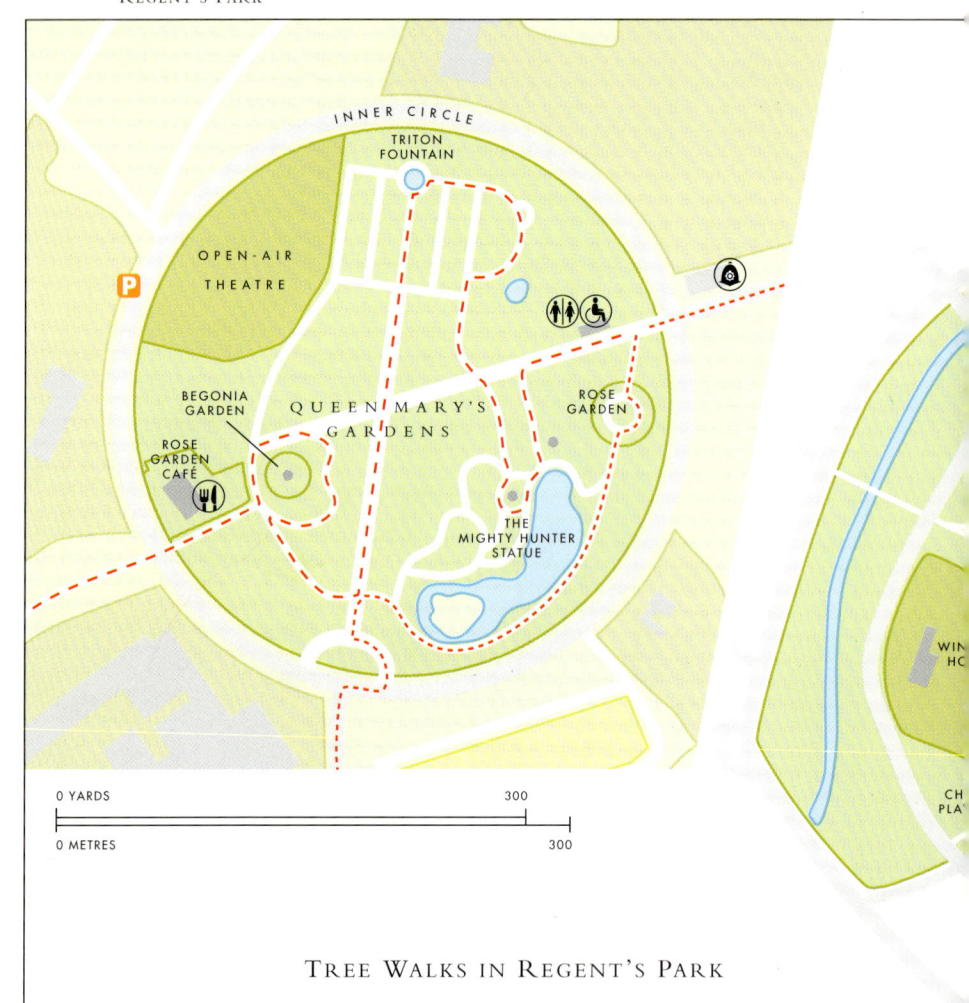

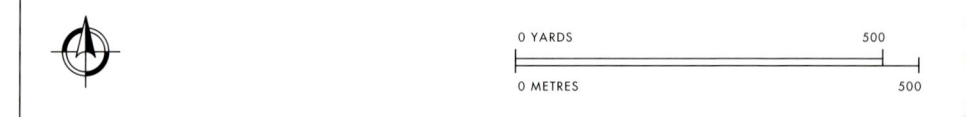

TREE WALKS IN REGENT'S PARK

Regent's Park

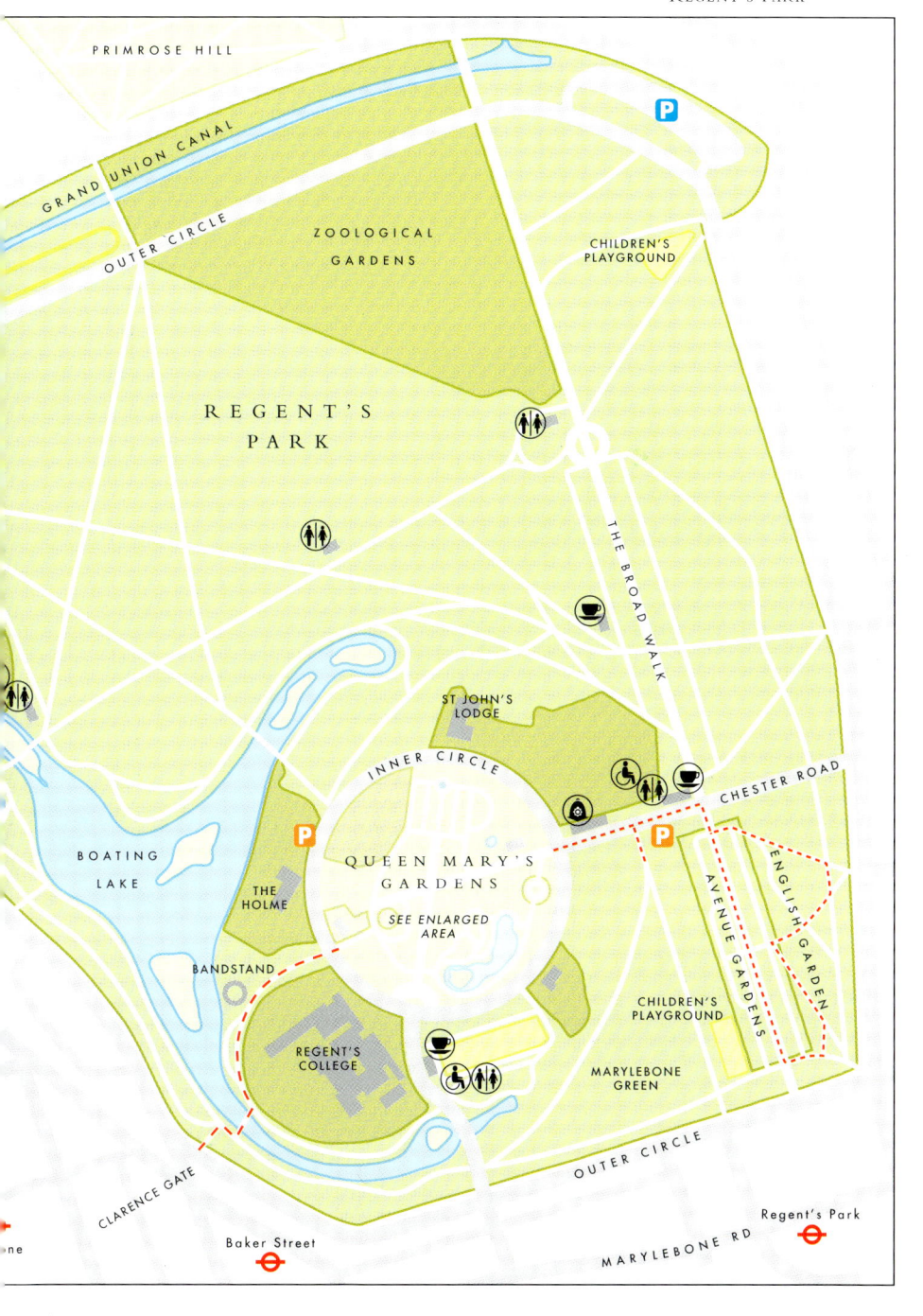

Primrose Hill

Grand Union Canal

Outer Circle

Zoological Gardens

Children's Playground

REGENT'S PARK

The Broad Walk

St John's Lodge

Inner Circle

Chester Road

Boating Lake

The Holme

QUEEN MARY'S GARDENS
See Enlarged Area

Bandstand

Regent's College

Avenue Gardens

English Garden

Children's Playground

Marylebone Green

Outer Circle

Clarence Gate

Baker Street

Marylebone Rd

Regent's Park

53

GREENWICH PARK

GREENWICH PARK

GREENWICH PARK, acquired in 1427 by Humphrey, Duke of Gloucester and youngest son of King Henry IV, is the oldest of the London Royal Parks. The land stretching from the Thames up the slopes to Blackheath was inherited by King Henry VI in 1447 and the old rambling Royal palace near the river, variously known as 'Placentia', 'Manor of Pleasaunce' and 'Bella Vista', eventually fell into disrepair and was demolished. Consequently, wherever digging is carried out in the northern part of the grounds, rubble, foundations, or made ground appears as soon as the surface is disturbed. In many places, the southern raised part of the site consists of sand, friable loams or gravels which overlie the bed of London clay which outcrops part-way up the scarp face; at one time, springs emerged from this point but only two conduit heads remain of the 14 that once existed.

Above the spring-line the topsoil is thin, drains rapidly and is easily eroded, so that trees large enough to catch the eye of great tree-measurers such as Alan Mitchell are few. Those clustered trees which straggle along the steeper slopes of the scarp approach woodland in status and visually divide the site into two: on the lower ground there is the open lawn near the Queen's House and above, on the plateau, the formal arrangement of avenues originally planted in the early 1660s.

The shallow banks with block-plantings of trees, which bear some kind of relationship with the Queen's House, are the only relics of a garden scheme prepared for Charles II by the great André le Notre (France's greatest garden designer and horticulturalist, employed by Louis XIV). The scaled plan, with the great man's handwriting on it, is still in existence in France. In the original

Facing page:
Common Magnolia
Magnolia x *soulangeana*
Royal Observatory
Garden

Right:
Sweet Chestnut trees
Castanea sativa
Great Cross Avenue

Left:
Prickly Castor-oil Tree
Kalopanax pictus

Facing page: Holm Oak
Quercus ilex

plan, the banks mark the edges of terraces while the present tree-plantings are replacements for the original block-plantings of the 1660s. These early plantings were still visible as ragged groupings in a survey of 1840 but, by 1897, had declined almost into non-existence.

These blocks were replanted in the early 1950s, at about the same time as the Broad Walk in Kensington Gardens. They consist of alternate lines of Common Beech (*Fagus sylvatica*) and Turkey Oak (*Quercus cerris*). Oak and beech compete furiously, and the oaks appear to be of mixed origin – they are definitely not of the same clone.

The formal scheme of avenues, most probably designed by André Mollet (one of the sons of Claude Mollet, Gardener to the King of France and the man from whom André le Notre learned the elements of gardening) have declined more than those of Kensington Gardens. The original lines of trees in the Great Cross Avenue have been increased to eight and, perhaps because of a lack of uniformity in gapping-up, and the damage done by 'desire-line' paths cutting across the symmetry, the avenue has a rather ragged appearance.

All the avenues have suffered to a greater or lesser extent. Originally, some were planted with elms and some with Sweet Chestnuts (*Castanea sativa*). A start has been made on the restoration of the avenues to something of their old regularity, using only the minimum of less appropriate species, such as horse-chestnut or plane, but the site is elevated and rather exposed, and tree growth on the plateau can be slow. Elms, for obvious reasons, are no longer a viable option.

The most impressive trees in the Park are the ancient Sweet Chestnuts – notably the one standing to the northeast of the Blackheath Gate, which still possesses some part of its upper crown. Even the almost total ruins, which can be found nearly all over the Park, are worth studying. The most interesting exotic species are mostly to be found in the Ranger's Field, the Flower Garden, or the recently planted (perhaps 40-year-old) garden on the western slopes of Observatory Hill.

A TREE WALK

Enter the Park through the **Blackheath Gate** and admire **Blackheath Avenue** – now planted with Horse-chestnuts (*Aesculus hippocastanum*). Even the inevitable 350 parked cars cannot remove its dignity. The Avenue has an eye-catching focal point at each end: the spire of All Saints Church on the other side of Blackheath and the imposing figure of General Wolfe gazing towards the Queen's House at the northern end.

Turn left, noticing the large old Tulip Tree (*Liriodendron tulipifera*) close to the public conveniences. Start your circumnavigation by walking fairly close to the perimeter wall and observe the extraordinary mixture of species, planted right around the boundary of what was once the enclosed, private garden for the Ranger of the Park. In this exposed place even common species find conditions difficult. There are a number of Sycamores (*Acer pseudoplatanus*), some of which show signs of decline, but oaks, both native and exotic, seem better able to flourish. Note the fine Common Oak (*Quercus robur*) in the small enclosure at the rear of the Police House. Hollies, mostly Highclere Holly (*Ilex* x *altaclerensis*), have been widely planted at some time, probably for wind shelter.

The young Sweet Gums (*Liquidambar styraciflua*) along the boundary are making promising growth and, near the **Ranger's House**, is a mature Holm Oak (*Quercus ilex*) and a middle-aged Red Oak (*Quercus rubra*), its bole encroaching on the pathway. In the **Rose Garden** is a cedar and a remarkably old and large Common Beech (*Fagus sylvatica*), remarkable, that is, considering the harsh environment. Near to the **tennis courts** is a touch of variegation, planted perhaps 20 years ago: ash cultivars, Red Oaks, Scarlet Oaks (*Quercus coccinea*), a middle-aged Indian Bean Tree (*Catalpa bignonioides*) and a Cappadocian Maple (*Acer cappadocicum*). Between the tennis courts and the perimeter planting around the circular **reservoir** is the first of the marvellous and gigantic ruins (relics even) of the ancient Sweet Chestnut (*Castanea sativa*) plantings – perhaps 3m (10ft) in diameter at the base.

The screen-planting around the reservoir includes Austrian Pines (*Pinus nigra* subsp. *nigra*) and Holm Oaks. On the eastern side of the enclosure are two flourishing Oriental Planes (*Platanus orientalis*); whatever else might be removed to restore the regularity of the avenues, these should remain because they are fairly rare. There is also a Deodar (*Cedrus deodara*) and a mature Red Oak. At the front of **Macartney House** is a False Acacia (*Robinia pseudoacacia*).

From Macartney House it is possible to look down the **Great Cross Avenue** and to see the ancient Sweet Chestnuts of the oldest plantings, as well as the many other species which were inserted in later centuries when the lines were gapped-up or extended laterally.

At the northern end of the ornamental border fronting Macartney House is a variegated cultivar of Tree Privet (*Ligustrum lucidum* 'Excelsum Superbum'), which, with any luck, will grow to three times its present height. A flourishing Turkey Oak (*Quercus cerris*) stands in the middle of the clear grassy space on the edge of the scarp and, about halfway to the **King George Street Gate**, is a tall Wild Cherry (*Prunus avium*), a fine sight in spring. From here walk over the slopes towards the base of the **Observatory Hill**, noting a prominent Blue Cedar (*Cedrus atlantica* f. *glauca*) in the copse partway down the scarp.

From the bottom of the Hill, walk up alongside the railings, observing the rich planting within the enclosure. About halfway to the junction with Blackheath Avenue, there is an opening in the fence and a path slopes steeply upwards towards the Observatory. Before entering look at the vigorously growing trees and shrubs on each side of the opening. The whole of this side of the Observatory Hill appears to have been very well planted with choice and rare species about 40 years ago and the larger plants are at, or approaching, maturity. On the left of the opening, near the fence, is a Flowering Dogwood (*Cornus florida*), glorious when it is in bloom, and a Fern-leaved Beech (*Fagus sylvatica* 'Asplenifolia') with elegantly snipped leaves – a vigorously

Ginkgo
Ginkgo biloba
Flower Garden

growing young plant which will eventually attain considerable size. Alongside is another Variegated Tree Privet and a thriving Southern Magnolia (*Magnolia grandiflora*).

On the other side of the entrance to the sloping pathway to **Flamsteed House** is a Cappadocian Maple (*Acer cappadocicum*) and a group of magnolias. Take this pathway and, halfway up, go through the entrance to the Royal Observatory Garden. On each side of the entrance gate, are specimens of Moosewood (*Acer pensylvanicum*), a 'snake-bark' maple. Look up from the gateway and, high above the Dome, you will see another Cappadocian Maple.

Turn left and, in the first angle of the Garden, you will see a group of Bladdernut Trees (*Staphylea colchica*); in spring, the individual flowers, which are borne in panicles, vaguely resemble small white daffodils. There is also a Chusan Palm (*Trachycarpus fortunei*) on the edge of the lawn, a Japanese Maple (*Acer* spp.) of tree size, and a healthy-looking Dove, or Handkerchief, Tree (*Davidia involucrata*) by the picnic seat. After a quiet rest to consult your tree identification book, continue up the sloping path. Those who persevere to the top of the slope will find, near the first buildings, an ancient fig tree with a notably pungent aroma. Continue past the **Meridian Line** and prospect point, and take the path which winds along the edge of the scarp in the direction of the **Maze Hill Gate**. The tree cover has the character of thin woodland. Some little way along, to the right, is a Red Oak, beyond which are some more ruined chestnuts with gargantuan, twisted and carbuncled boles. They are at the end of one of the original diagonal avenues, planted in the early 1660s, which go down the slopes of the scarp, not far from **Queen Elizabeth's Oak**. This oak is the recumbent relic of a hollow tree (now surrounded by railings) in which, at one time, a banqueting party could be held. There is an actively growing replacement planted inside the enclosing fence.

There follows an area planted with all kinds of oak species, amid a confusion of little dry valleys. One of them, **Lovers' Walk**, has been replanted with hornbeams wherever possible. Near the **Roman Remains**, you can stand beneath the canopy of an antique but healthy Common Oak (*Quercus robur*) that has a tremendous girth. However, if you stand back and consider the size of the crown it will be seen as diminutive when compared with the massive base. It is highly probably that, because of the harsh conditions, this tree, like many others in the Park has never had a crown properly proportionate to the size of its supporting trunk.

Now walk towards the **Flower Garden**. On the way are many monumental chestnut wrecks, all very sketchable. Enter the Flower Garden where it impinges upon the Great Cross Avenue. This Garden has a 'Victorian

Left: Paper-bark Birch
Betula papyrifera

Above: Common Beech
Fagus sylvatica
Ranger's House

One-leaved Ash
Fraxinus excelsior
'Diversifolia'

Afternoon' atmosphere of its own. Inside the insulation of the enclosing evergreen hedges and the borders are well-spaced-out cedars, mainly of that essentially Victorian villa garden species, the Deodar, and a few pines, such as the Scots Pine (*Pinus sylvestris*) and the Bhutan Pine (*Pinus wallichiana*). Contrasting with the dark conifers are some middle-aged Tulip Trees, a few walnuts, some mulberries and a One-leaved Ash (*Fraxinus excelsior* 'Diversifolia').

There are a number of Ash-leaved Maples (*Acer negundo*) in the borders. These were probably one of this species' spectacularly variegated cultivars when planted, but, unless fanatically cared for, they invariably revert to green and then grow very quickly! In the border adjacent to the **Wilderness** is a Persian Ironwood (*Parrotia persica*). However, the great rarity of the Garden is the small group of Prickly Castor-oil Trees (*Kalopanax pictus*) in the northern border. This highly unusual species, of great character, with shallowly lobed leaves, has a quite extraordinary appearance when it flowers in late summer: dead-white flowers and flower stalks, borne in umbels like little cartwheels 20cm (8in) wide, appear in abundance all over the crown.

From here stroll around to the nearest **Dell**, of which there are two in the Park. This area boasts a huge Ginkgo (*Ginkgo biloba*), a Californian Laurel, or Headache Tree (*Umbellularia californica*) and, near the lake, a *Pittosporum tenuifolium* and a Nikko Maple (*Acer nikoense*), wonderful in its autumn colour. In the middle of the **Heather Garden** is an isolated specimen of the most beautiful cultivar of the Common Alder (*Alnus glutinosa* 'Imperialis'), and, near the Dell entrance, a Shagbark Hickory (*Carya ovata*) and a Tupelo (*Nyssa sylvatica*).

By the lake margin is a Swamp Cypress (*Taxodium distichum*), a Dawn Redwood (*Metasequoia glyptostroboides*), a large White Willow (*Salix alba*), an Indian Bean Tree (*Catalpa bignonioides*) and a large Flowering Dogwood. Make your way out past an infant Antarctic Beech (*Nothofagus antarctica*) and, even if your head is starting to spin, before leaving admire one of the Grandees of the Park – the veteran chestnut near the Ranger's Lodge, which still carries a large part of its crown.

At this point, you are not far from the Blackheath Lodge Gate and the end of the walk.

Sweet Chestnut
Castanea sativa

GREENWICH PARK

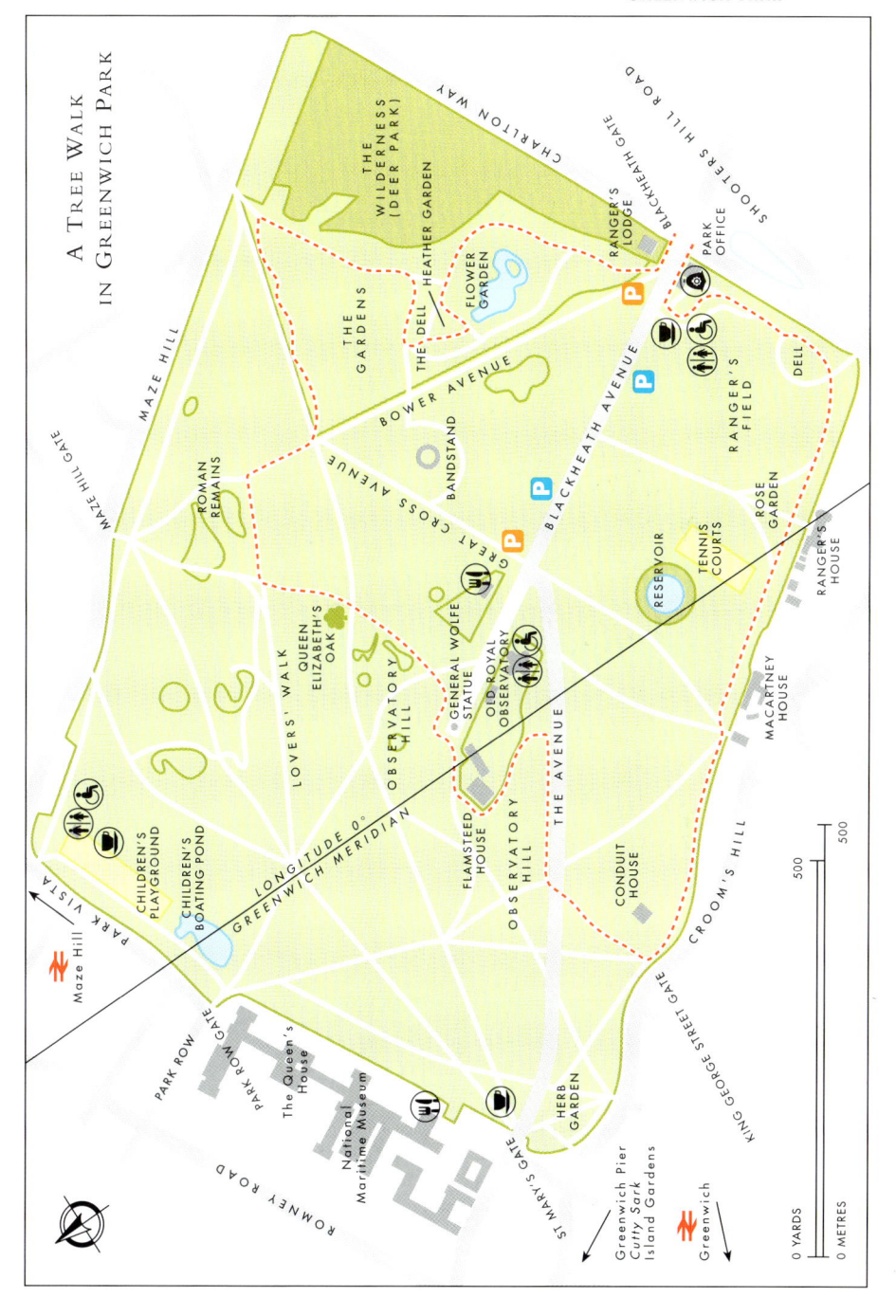

A TREE WALK IN GREENWICH PARK

Richmond Park

RICHMOND PARK

THIS PARK was a playground for Royalty long before King Charles I acquired it in 1637, enclosed it, and named it 'Richmond Park'. It lies close to the former Richmond Palace (originally Shene Palace), which has been owned by the ruling monarch since the early Middle Ages, and the whole area was hunted over for centuries.

King Henry VIII's Mount, which stands between Pembroke Lodge and the Richmond Gate, is the point where the King is supposed to have stood to shoot driven deer and it is still a fine vantage point. If you turn your back on the view and put your eye to the little aperture in the screen hedge, you will see, 16 kilometres (10 miles) away, as if through a telescope, St Paul's Cathedral at the end of a long, narrow alley cut through Sidmouth Wood.

The Park's long history of use for deer-hunting has been of great benefit to those trees which were already standing when it was enclosed. Since time immemorial they have been a resource for local farmers, who pollarded many of them to provide fuel, wood for small buildings and furniture, bark for tanning, and browse for their animals. The trees would thus receive a periodical 'haircut', above easy biting height for the animals. Ancient pollard and standard oaks of immense size exist here and there all over the Park, some in the open, but many enclosed in woodland. It is impossible to assess their age from annual growth rings because trees of such age do not lay down a regular increment of growth each year; indeed, some stop their secondary thickening

Facing page:
Common Hornbeam
Carpinus betulus
Hornbeam Walk

Right:
Ancient Oaks
Quercus robur
High Wood

almost entirely. Pollarding and coppicing appear to interfere with the onset of old age – as well as seeming to prolong active life.

Ancient trees are not too uncommon in Britain, but communities of this size are rare in the British Isles, and indeed exceptional in continental Europe. The site, its trees, and the hosts of animals and fungi dependent on the old timber, both living and dead, are of great scientific importance. Consequently the Park has been declared a Site of Special Scientific Interest and its maintenance is now subject to an overview by English Nature. This, however, has not seriously interfered with the growing of useful timber and its conversion, on site, into timber products. Open Days are arranged from time to time, usually in October, when timber extraction by a team of heavy horses, and timber-handling, can be seen.

Probably the Park's most enthusiastic tree-planter was Viscount Sidmouth, who was involved with the Park from 1801 to 1844. It was he who enclosed the Isabella Plantation and planted Sidmouth Wood and several other smaller copses, using mainly indigenous species, with a preponderance of oak. He had a great eye for harmonious design in the arrangement and shaping of tree masses, and the contrasting open spaces between them, and showed great taste in incorporating existing ancient trees into his designs. After more than 150 years have elapsed, we are now able to appreciate those plantings in their maturity.

Nevertheless a veteran arborist, stated that, when he first saw the Park, from the back of a military vehicle in 1943, he was horrified by the condition of the trees: there appeared to have been little new planting, the woodland seemed to be in decline from lack of thinning, and stag-headed oaks appeared to predominate. The next time he viewed it, in 1960, the scene had been transformed by George Thompson, a professional forester. As well as reputedly planting more than 70,000 trees during his Superintendency (1951–1971), Thompson also developed the vastly popular horticultural underplanting of the Isabella Plantation.

With a beady eye on the future, one of George Thompson's practices was to plant three sapling oaks, widely spaced around veteran standards and pollards. Eventually, the old trees will decay but, as they fall apart, the young trees can be singled out to take their place, or new pollards made. An additional programme of pollarding the waterside willows is now being instituted, under the close supervision of the present Managers.

Probably, the most visited tree in the Park is John Martin's Oak. Martin was a painter in the middle of the Victorian period, who specialised in disaster scenes – Babylon glowing red-hot, with highly enjoyable incandescent bombs and fountains of flame and fume comes to mind. What possessed him to paint a paradisiacal picture of a grandfather oak is a matter for conjecture.

John Martin's Oak
Quercus robur

Nevertheless, just such a picture can be seen in the Ellison Gift of the Victoria and Albert Museum, and the tree itself, hardly changed, can be found at the southern end of the Hornbeam Walk. This tree demonstrates the life history of the oak – 300 years to grow, 300 years to exist in maturity, 300 years to die.

A curiosity, standing between Adam's Pond and Sheen Cross Wood, is the small, iron-fenced enclosure that used to protect the remains of the Shrew Ash, which was once believed to have curative properties. In early times, a shrew would be placed in a hole bored into an ash tree and the hole would be securely plugged. (In the case of this tree, old pictures show that the trunk had been split and wedged open with a bar of wood, about 2m (6ft) from the ground.) Any paralysis in domestic animals was thought to be the result of contact with a shrew and the remedy was to obtain a twig from a shrew ash and pass this over the affected limbs. Sick children were brought to the tree at sunrise, in the care of a 'Shrew Mother', and passed nine times around the tree while doggerel verses were recited or sung. A replacement tree has been planted to keep alive the memory of these folk remedies.

Tree Walks

It is possible to walk all the way around this very large open space, if properly clad, shod and provisioned, but no such route will be recommended here. There are car parks, and entrances near to public transport (see map p.75), and so only a few short walks to the most outstanding areas are described.

Broomfield Hill Car Park to the Isabella Plantation

From the entrance to Broomfield Hill Car Park, look across the road to the fenced enclosure that contains only conifers. This is the only such planting in the Park and it is intended to block the view of traffic on the roads in the middle distance and also to variegate the colours of the otherwise fairly uniform copses and spinneys. A dramatic dark mass on the skyline will eventually be formed as the plants mature.

Take the path which leads past the **Broomfield Hill Gate** and follow the boundary of the **Isabella Plantation** to the left. This area is known as **High Wood** and contains some of the finest high pollards and ancient ruined trees. Situated here is the **Deer Sanctuary**, which not only affords the deer a bit of peace but also protects the ancient pollarded oaks inside from the heavy treading which can severely damage old trees.

Enter the Isabella Plantation by the **Bottom Gate**, close to the disabled persons' parking place. Inside the gate is **Peg's Pond**, in which is a small island (**Wally's Island**) named after the Foreman Gardener, Walter Miller, whose enthusiastic and highly skilled gardening did so much to make the plantings a success during George Thompson's tenure.

Around the pond and on the island are a variety of willows. The long, crowded, bright golden boughs of the Golden Weeping Willow (*Salix* 'Chrysocoma') offer a pleasing contrast with the shorter, less highly coloured twigs and twig masses of the Babylon Willow (*Salix* x *sepulcralis*) and the pale leaves of the White Willow (*Salix alba*). The feathery conifer on the bank is a Swamp Cypress (*Taxodium distichum*). When the development of the plantation began in 1950, the tree canopy was lightened by felling and limb removal but there has never been any move towards changing the overall dominance of Common Oak (*Quercus robur*). In order to maintain the character of the oak woodland, only a small number of large exotic species have been planted and only a few of these show through the boundary of the plantation. Most of the more interesting exotic trees are small.

Walk straight along the main pathway beside the watercourse. To the right is a metal plate, dated 1831, which denotes when this part of the woodland was enclosed. The nearest oak tree is obviously too young to have been planted at that time. In this central area, some of the rhododendrons and camellias are of tree size.

A little way along the path, by a bridge, at a point where the paths bend to the left, is a small clearing on the right; in the middle is an Antarctic Beech (*Nothofagus antarctica*), with the small, regular leaves typical of the species. You are now in a glade which leads circuitously towards the Broomfield Hill Gate. A short way up is a tall, dark, evergreen conifer, a Lawson Cypress (*Chamaecyparis lawsoniana*), beyond which is a Common Beech (*Fagus sylvatica*), leaning heavily over with its root system exposed. Now return to the main path.

Continue towards the gate and inspect the two large conifers nearby: the one with feathery deciduous leaves is a Dawn Redwood (*Metasequoia glyptostroboides*) while the other, darker tree, more heavily furnished with foliage, is a Lawson Cypress. Retrace your steps for about 60 paces from the conifers to a small clearing on the northern side of the path which surrounds a small, very rare evergreen tree – *Euonymus myrianthus* – one of the spindles and, like many of the genus, spectacular in autumn, when in fruit. Carry on past the clearing towards the large, dark Austrian Pine (*Pinus nigra* subsp. *nigra*) and after inspecting the strongly growing young Tulip Tree (*Liriodendron tulipifera*), turn left and walk on towards the **Still Pond**, past the toilet block.

Snowbell Tree
Styrax japonica
Isabella Plantation

Left: Hollow Oak
Pembroke Lodge

Facing page:
Ancient Oak
Quercus robur
near Holly Lodge

Where the path forks, look out for small trees on each side of the path. To the right is a Snowbell Tree (*Styrax japonica*), which bears exquisite, pendulous white bells in spring; it will probably double in height as it matures. To the left, a few metres from the path among the rhododendrons, is a Deciduous Camellia (*Stuartia pseudocamellia*). Its bark, which flakes in patches, becomes more striking as the tree matures; the blossoms resemble small, single, off-white camellia flowers and the leaves may turn bronzy-red in the autumn.

Take the right-hand path towards Still Pond and walk clockwise around it until you reach the centre of the straight western bank. Face directly away from the water and you will be looking straight at an ancient beech tree, about 35m (40yds) away, with huge branches originating from the main trunk at less than a metre from the ground. The beech's rugged arms enclose a sturdy oak tree which it has 'married'!

Return to the path near the boundary of the Plantation and proceed towards the **Deer Sanctuary Gate**. Along this stretch of the perimeter, the indigenous trees and bushes are closing in – among them, alongside this minor path are specimens of native Bird Cherry (*Prunus padus*), with pale brown bark peeling off in thin horizontal strips. This species is not native locally, so the original trees must have been planted, although it is now self-seeding. The leaf stalks lack the conspicuous glands found in the Wild Cherry (*Prunus avium*) and its white flowers are borne in racemes. This area, which contains many large and eccentrically shaped trees, is one of the least altered parts of the Plantation and contrasts strongly with the central ones of brightly coloured exotic plantings.

Continue down the perimeter path, cross the central stream near Peg's Pond and make your way towards **George Thompson's Pond**. This more recently developed part of the Plantation has some very promising young trees. Near the pond is that recent (in gardener's time) incomer from China, the Dawn Redwood, with feathery, deciduous foliage and fine, fiery autumn colour. If the side branches are removed from a young plant as it grows, the lower stem will form a fairly smooth, rounded bole; on the other hand, if the side branches are left intact, the bole develops deep eyes and convolutions of great character. This tree is developing an interestingly gnarled trunk. Towards the northern end of the little lawn is a young Tibetan Cherry (*Prunus serrula*) with highly polished bark and, in spring, small white flowers.

Just before crossing the bridge onto the main lawn, take a look at the Snowdrop Tree (*Halesia carolina*), growing not far from the path almost 20m (22yds) further along. With its masses of pendant white flowers, this is glorious in spring. Just behind it is a young Swedish Birch (*Betula pendula* 'Dalecarlica'), with handsomely cut leaves. On the other side of the pond, near the exit from the lawn, is a tree of pyramidal shape, clothed all the way down to the ground. This, a Tupelo Tree (*Nyssa sylvatica*), has startling orange autumn colour and tends, as it ages, to lean on its 'elbows'. Old trees at Sheffield Park and nearby Cannizaro Park show this feature, which contributes to the tree's stability, but this tree is, as yet, only leaning on the tips of its 'fingers'.

From the lawn, proceed towards the main gate. To the right are two conifers lifting their pointing fingers through the oak canopy: the immensely tall Giant Sequoia (*Sequoiadendron giganteum*), with its characteristic, thick, spongy bark, is only a young plant and has attained only half of its possible eventual height; the other is a Coastal Redwood (*Sequoia sempervirens*), a close relative.

The walk ends at the Broomfield Hill Gate.

PEMBROKE LODGE TO JOHN MARTIN'S OAK

The grounds of **Pembroke Lodge** contain a number of characterful oaks (especially the prominent hollow one near the house, which is a favourite children's hidey-hole) and thriving evergreen Southern Magnolias (*Magnolia grandiflora*) on the garden front of the house. The most interesting species, however, is to be found near the extreme southern end of the garden, among the hollies and bays screening the adjacent car park from the enclosed garden. Here emerges an evergreen tree of pyramidal form with upwardly pointing pale leaves – an Australian Sassafras (*Atherosperma moschatum*). This tree is far from its native Australia and is frost-tender, requiring a minimum winter temperature of 3–5°C (37–41°F). The Pembroke Lodge grounds stand on the edge of a scarp and are open to gales from the southwest but, nevertheless, the tree appears to be healthy. It bears small white flowers in spring and has nutmeg-scented leaves but these are far out of reach. Long may it persist.

Look at the ancient pollards in the enclosure. Many of them have been pollarded at above the usual height, because they were intended to be nesting and roosting places for the turkeys which were kept here in Tudor times, not longer after their introduction to England.

Turkeys, heresy, hops and beer
Came into England all in one year.
Foxes think little of scaling a 2m (6ft) wall but one imagines that the turkeys would have been pretty safe on top of these pollards.

Leave by the car park gate and turn right under the canopy of the **Hornbeam Walk**. This decayed avenue of Hornbeams (*Carpinus betulus*) is unlikely to be reconstituted because, even when it was planted, it could only have been intermittent, owing to its interruption by trees already growing on the site. Some of the hornbeams, with their twisted, snakeskin-patterned trunks, appear to be of great age. Proceed along the edge of the scarp as far as the road to Ham Common. Just before this point are a number of huge oaks, all worthy of notice. Consult the photograph to identify which is **John Martin's Oak**. The walk ends beneath its branches.

Real enthusiasts will also walk from the **Richmond Gate**, down **Queen's Ride** to **White Lodge**. It is a pleasant place to saunter and a good spot in which to catch a glimpse of the green parakeets which haunt the quieter parts of Richmond Park.

Oak/Beech 'marriage'
Isabella Plantation

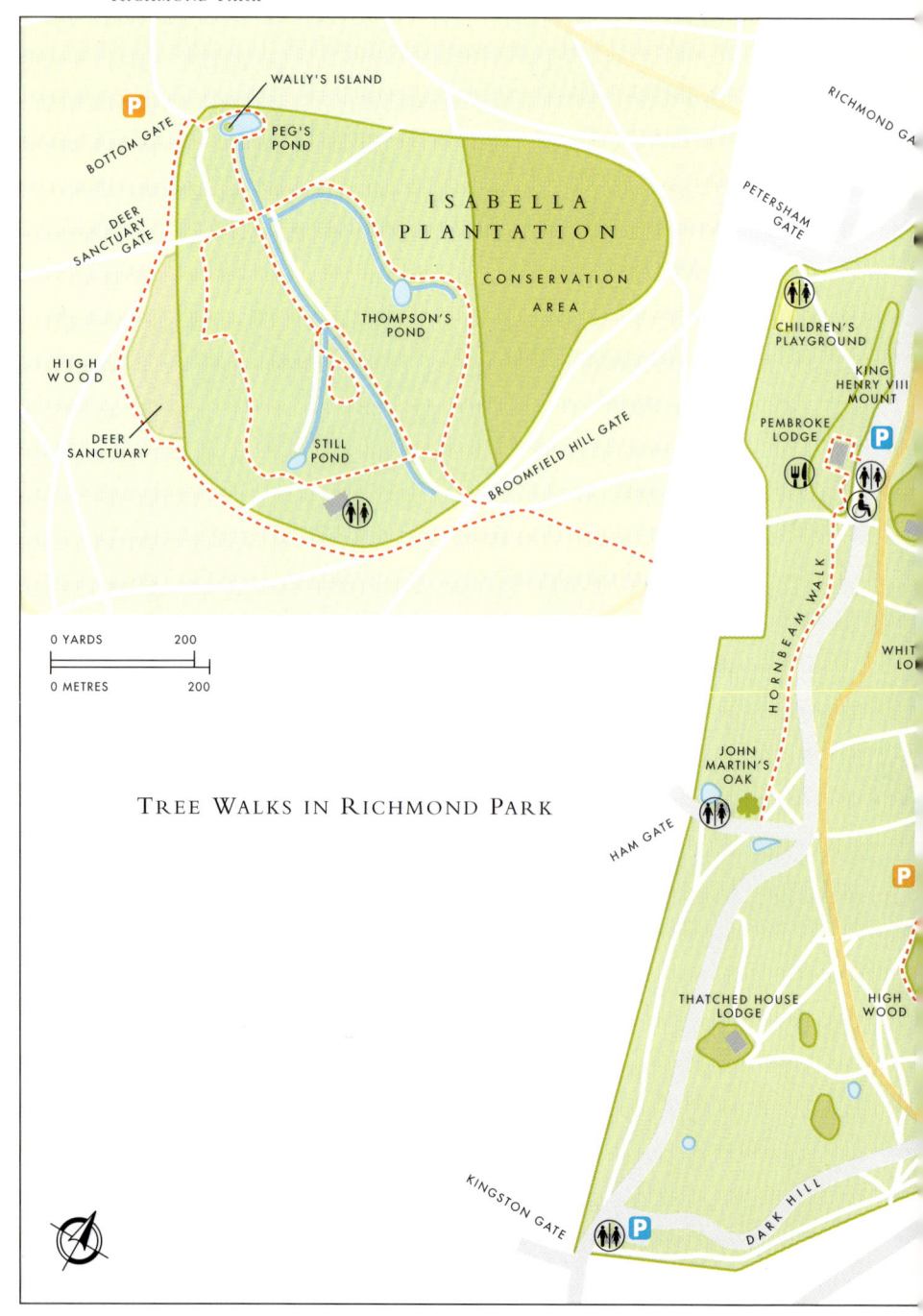

RICHMOND PARK

TREE WALKS IN RICHMOND PARK

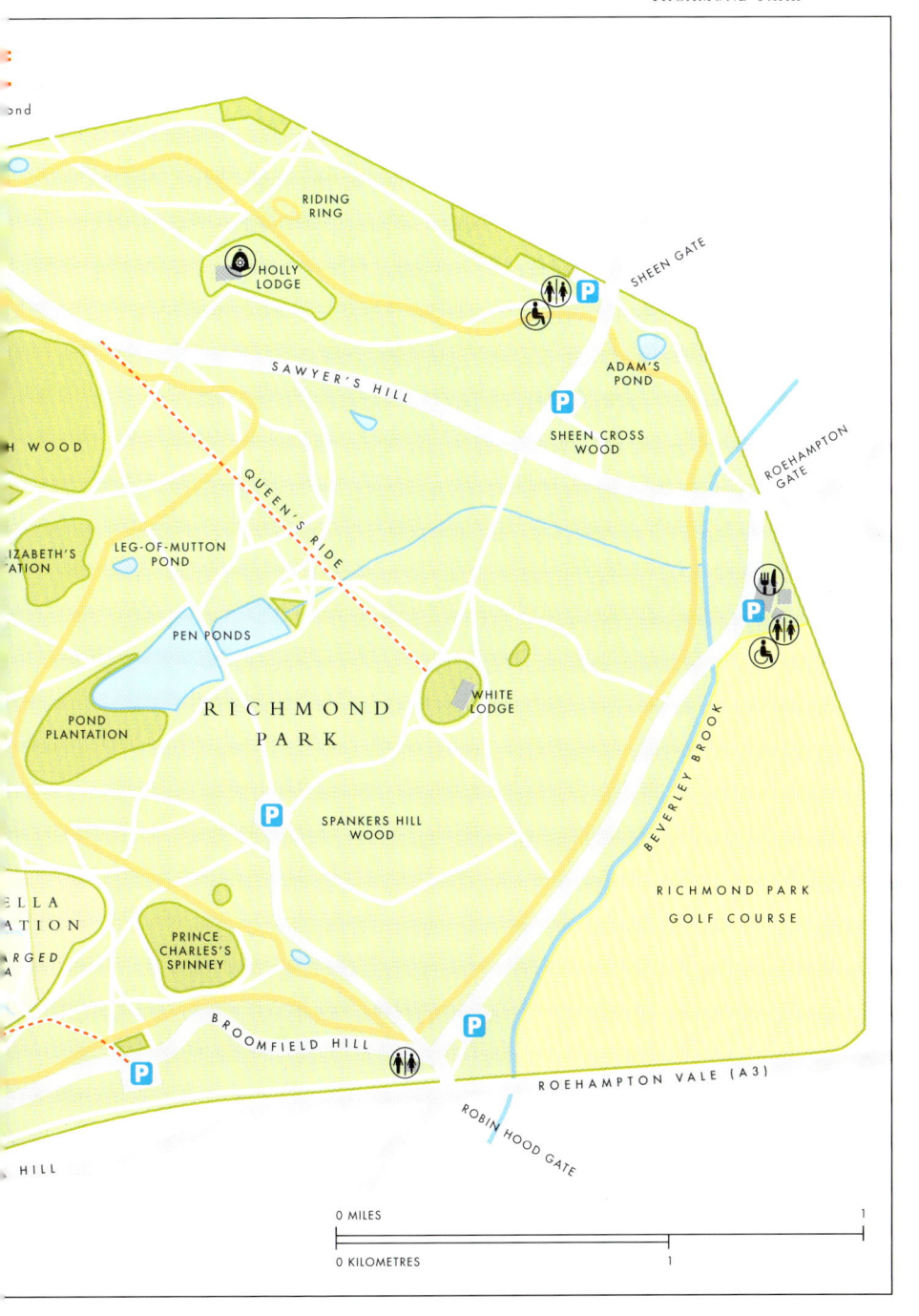

RICHMOND PARK

RIDING RING
HOLLY LODGE
SHEEN GATE
SAWYER'S HILL
ADAM'S POND
SHEEN CROSS WOOD
ROEHAMPTON GATE
QUEEN'S RIDE
LEG-OF-MUTTON POND
PEN PONDS
WHITE LODGE
POND PLANTATION
RICHMOND PARK
BEVERLEY BROOK
SPANKERS HILL WOOD
RICHMOND PARK GOLF COURSE
PRINCE CHARLES'S SPINNEY
BROOMFIELD HILL
ROEHAMPTON VALE (A3)
ROBIN HOOD GATE

0 MILES 1
0 KILOMETRES 1

BUSHY PARK

BUSHY PARK

THIS PARK has retained more of the character of a hunting-ground than any of the other Royal Parks in the London area. Richmond Park may be a splendid piece of landscape but, with half of it given over to woodland, it is too cluttered to be a pleasant place to hunt nowadays. Bushy Park, however, is still largely open and, even today, could give a huntsman a good scramble.

At the times when the Kings and Queens of England, and other great land-owners, decided to extend their hunting-grounds, it obviously made sense for them to take over impoverished estates, with poor-quality soil, supporting meagre plant growth, and impeded drainage. Today, such land is found to consist of infertile, light sandy loams, which are acidic and poorly drained. These conditions, as every budding horticulturist will recognise, are perfect for growing shrubs such as rhododendrons, camellias and magnolias. Thus we have inherited those two jewels: the Waterhouse Plantations (now the Woodland Garden) in Bushy Park and the Isabella Plantation in Richmond Park.

The glories of Bushy Park are: the Great Chestnut Avenue, which is somewhat of a misnomer because only the middle pair of lines are horse-chestnuts, the other lines being made up of limes; and the Diana Fountain. This multiple avenue has always been effectively maintained – decrepit trees being removed

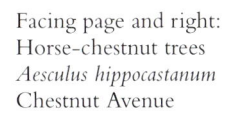

Facing page and right:
Horse-chestnut trees
Aesculus hippocastanum
Chestnut Avenue

before they fell over and replaced with the same species – so complete renewal has never been necessary. Perspective and distance will conceal any occasional temporary unevenness. However, the arrangement of trees in 'great bows' or 'rounds', which are viewed from the front, are more problematical; severe storms may so damage old trees that complete renewal may be necessary. This has happened to the trees surrounding the Diana Fountain.

The other significant formal feature of Bushy Park is the Lime Avenue, which suffered extensive damage in the 1987 hurricane, when 148 trees were lost in this avenue alone, but has since been restored. It was once a cross-avenue (there is a cropped limb on the eastern side of the Diana Fountain) but, whereas the Chestnut Avenue has no outstanding feature at the Teddington Gate, where it ends, the Lime Avenue has a focal point at both ends. To the west is the pleasant Georgian architecture of the White Lodge, which can be viewed from the Diana Fountain.

Apart from these two avenues, there is no sense of overall design or unity in the Park. The informal water features, almost all artificial, are pleasant enough but the fragments of avenues and of formal water schemes, leading in no logical direction, are baffling. There appears to be an uncompleted avenue centred on the Upper Lodge and running up from Waterhouse Pond, a transverse, more complete avenue along the bottom of the Upper Lodge garden and, on the other side of the Lodge, a truncated formal canal.

The oddest relic, well worth inspecting, is a tiny and sinisterly dark avenue of Fastigiate Hornbeams (*Carpinus betulus* 'Fastigiata') near the River Lodge Gate. Here, the crowns of the trees have coalesced to form a complete canopy.

Lime Avenue
Tilia x *europaea*

Right:
Magnolia 'Kewensis'
Woodland Garden

Below: Snow Gum
Eucalyptus niphophila
Heather Garden

This form of hornbeam is slender and graceful in youth, obese in middle age and, in old age, develops a crown with rigidly straight branches, which radiate from the top of the trunk like bristles in a lavatory brush.

The trees in the open parkland and in the smaller fenced enclosures are mainly pleasant native species, such as oak, ash, beech, lime, but they have been somewhat restricted in size by the poor quality of the soil. The more unusual tree species are to be found in the Woodland Garden. The horticultural underplanting of these spaces was begun by Superintendent Joseph Fisher, in 1948/9, shortly before his retirement. The work was carried on by his successor, George Cooke, and its maintenance and variegation have been continued up to the present day.

A Tree Walk

Enter by the **Crocodile Gate** into the **Woodland Garden**. On the far side of the stream there is a cedar, a young and burgeoning Holm Oak (*Quercus ilex*) and a fine group of Common Oaks (*Quercus robur*) and, in the open glade beyond, a mixture of trees that existed prior to the horticultural underplanting, surrounded by more recently planted saplings. Two very promising Monterey Pines (*Pinus radiata*), with needles characteristically grouped in clusters of three, stand in the screen-planting to the rear of the cedar. In the middle of the more open ground is an old Sweet Chestnut (*Castanea sativa*) and, among a scattering of young trees, a promising Coastal Redwood (*Sequoia sempervirens*) and a Keaki (*Zelkova serrata*), a relative of the elm.

Return to the main path and continue for a short distance. In the southern boundary border, to the left of the path, is a Caucasian Wingnut (*Pterocarya fraxinifolia*) with a mass of suckers on its roots, spreading in all directions, and opposite, on the far side of the stream, are two splendid London Planes (*Platanus x acerifolia*), exotic trees in this setting. On the near bank, alongside the water, are two mature Swamp Cypresses (*Taxodium distichum*); both have prolific, knee-like growths (pneumatophores) on their roots and these protrude from the water close to the bank. These act as breathing organs and enable the trees to flourish in swamps and standing water where other species would perish. There is also an immature Swamp Cypress, planted for the future, and more specimens of the same species on the far bank.

A little further along the path, where the view to your half-right is clear, the biggest and most handsome Swamp Cypress may be seen on the far bank. At this point, on both sides of the path, near the first small pond and island, are several Cappadocian Maples (*Acer cappadocicum*), with simple palmately lobed leaves and, in good years, fiery autumn colour. To the rear of the group

Left: Roble Beech
Nothofagus obliqua
Woodland Garden

Facing page:
Swamp Cypress
Taxodium distichum
Woodland Garden

is a single specimen of the Paper-bark Maple (*Acer griseum*), with its lovely, flaking orange bark.

Continue along the path until the space to the left opens out into a small glade. This is occupied by some pines in young middle age. The nearest to the approach, with foxy-red bark in the upper crowns, are Scots Pines (*Pinus sylvestris*), with paired needles. The row of handsome blue conifers are Bigcone Pines (*Pinus coulteri*), with needles in groups of three; look up to the top of the crowns to see the constellations of huge cones. The final lines are of Weymouth Pine (*Pinus strobus*), with needles in groups of five. Here again, it is rewarding to look at the upper crowns against the sky – the fan-shaped leaf bundles make a fascinating pattern against the light. Close to the path is a solitary, densely foliaged Monterey Pine (*Pinus radiata*). A little further along, growing close to the fence line, is a row of Serbian Spruce (*Picea omorika*), with typical, deeply recurving branches.

There are more Swamp Cypresses by the largest pond (**Triss's Pond**) and, to the left of the path, a Blue Cedar (*Cedrus atlantica* f. *glauca*). From this point, the path weaves among mature oaks and rhododendrons in a very satis-

Far left: Pin Oak
Quercus palustris
Woodland Garden

Left: Jacquemont's Birch
Betula utilis
var. *jacquemontii*
Woodland Garden

factorily mazy way. Opposite the **Woodland Cottage** is a small stand of vigorously growing Turkey Oaks (*Quercus cerris*) and, near the end of this section, with the gate in sight, a Roble Beech (*Nothofagus obliqua*) can be seen to the left. Just before the gate, also to the left, there is a sapling oak with enormous leaves, which resemble those of a hugely enlarged English oak; this is a Japanese species, the Daimyo Oak (*Quercus dentata*).

Pass through the gate and across the open space but, before entering the next gate, pause to look at the climbers scrambling among the trees that overhang the boundary fence, south of the watercourse. These are specimens of the twining shrub, *Celastrus orbiculatus*. The one nearest the gate is a male plant, while that furthest away is female. In autumn, the female bears quaint, lobed, dull orange fruit capsules which, like those of the related Spindle Tree (*Euonymus* spp.), suddenly flip open to reveal their bright red fruit within. A little further along the fence is a Golden Rain Tree (*Koelreuteria paniculata*), which can be seen more clearly from outside the enclosure.

Enter the gate into the next part of the garden, which is more richly planted. Almost immediately to the right of the gate is a surprisingly tall specimen of what is usually regarded as a dwarf cultivar ('Fletcheri') of Lawson Cypress (*Chamaecyparis lawsoniana*). The size of this is an object lesson in the interpretation of plant catalogues! Just beyond is an evergreen Eucryphia (*Eucryphia* x *intermedia*) which, like other members of this genus, is a delight in late summer, when it is usually covered with white-petalled flowers, each with a fuzzy boss of golden stamens. A few metres further along, in the strip of lawn, there is a Ginkgo (*Ginkgo biloba*) and, a little further still, another

Eucryphia (*Eucryphia glutinosa*), sunk in the shrubbery; this is a marvellous small tree whose leaves blaze with autumn colour before they fall. Turn left, cross the stream, and walk down through the leafy green tunnel towards the pond. There are Common Alders (*Alnus glutinosa*) here and there throughout the Garden – some old enough to pre-date its development by at least a century and one large enough to provide timber for about 100,000 clog soles. Near the first, smaller bridge are two more Swamp Cypresses and to the right, on the approach to the larger bridge, is a fine Common Oak.

By the pond – near a large beech – is a bushy tree with flaking bark, brown velvety buds and broad, serrated leaves, like a hazel, which also develops fine autumn colour. This is a Persian Ironwood (*Parrotia persica*), which starts off as a bush but develops into a characterful tree as it slowly matures. At 70 years old, it may be 10m (33ft) high, and can eventually attain about 15m (50ft). There are fine poplars to the east of the pond and several Swamp Cypresses on its margins. Downstream of the bridge, on the northern bank, is a spreading Italian Alder (*Alnus cordata*).

Cross over to the **Heather Garden** (**Fisher's Field**). The Snow Gum (*Eucalyptus niphophila*), near the opposite corner, has conspicuous, pale, shining bark and twinkling, pendent leaves and, in the island bed, there is a fine *Magnolia* 'Kewensis', with masses of upturned white flowers on naked branches in mid-spring. Alongside, in the border, is the less often seen Downy Tree of Heaven (*Ailanthus vilmoriniana*), whose leaves lack the lobes and glands found in the common species (*Ailanthus altissima*). The blue-leaved form of Scots Pine found in this area is outstanding and, on the edges of the lawn, are

Golden Weeping Willow
Salix 'Chrysocoma'
Woodland Garden

Horse-chestnut trees
Aesculus hippocastanatum
Chestnut Avenue

forms of the Southern Magnolia (*Magnolia grandiflora*), showing different densities of bright-brown woolly hairs (indumentum) under the leaves and varying intensities of flowering. There is a fine Medlar (*Mespilus germanica*) a short distance away, and stands of Western Red Cedar (*Thuja plicata*) and Lawson Cypress; the Western Red Cedar has a heavenly aroma of pineapple, while the Lawson Cypress smells of parsley. Both are marvellously healthy plantings. Nearby are specimens of Winter-flowering Cherry (*Prunus subhirtella* 'Autumnalis Rosea'), a cultivar which is also scattered elsewhere about the Garden. Before leaving the Heather Garden take a look at the beautiful Colorado Blue Spruce (*Picea pungens* var. *glauca*) to the left of the bridge.

Cross the bridge to **Lightning Beech Glade**. Just past the very tall alder is a labelled Pin Oak (*Quercus palustris*), a wet-ground species of American Oak. (Oaks like this, with pointed-lobed leaves can be difficult to identify.) Circumnavigate the large open area by **White Lodge**, join the path accompanying the stream, and walk towards **Waterhouse Pond**. In the corner by the gate, making good progress, is a Dawn Redwood (*Metasequoia glyptostroboides*). About halfway to the pond is a mysterious ruined brick shell. What was it? A grotto? Or the original waterhouse?

Alongside the pond, behind the vantage point, are some fine ash trees and a Silver Maple (*Acer saccharinum*) stands on the northeastern bank. Another fine ash grows alongside the northeastward-flowing stream, and there is a Dawn Redwood on the opposite bank. The excellent white-stemmed birch beside it is a Jacquemont's Birch (*Betula utilis* var. *jacquemontii*). Near the path is an evergreen Eucryphia (*Eucryphia x nymansenis* 'Nymensay'); both this and the deciduous *E. glutinosa* can be found on the northern perimeter path.

Pass through the **Sub-tropical Garden**, where there are recently planted Chusan Palms (*Trachycarpus fortunei*) and Dwarf Fan Palms (*Chamaerops humilis*), as well as forms of the Cabbage Tree (*Cordyline australis*). The Cabbage Tree is a fairly tender species and may need replanting after the next severe winter; the palms on the other hand, are rock-hardy and, when mature, may even flower in warm years. Make your way back towards the entrance, taking a detour to the largest evergreen Eucryphia, a superb plant when in bloom in late summer. This same path affords an opportunity to see a large group of mature evergreen Eucryphia (probably *Eucryphia* x *nymansensis* 'Nymansay') a little further along. Continue along the path, passing by some ancient Crack Willows (*Salix fragilis*) – the specific name is highly appropriate because the twigs crack off easily and the branches are also liable to break.

At the end of the path is the gate leading into the **Ash Walk** by which you entered.

Right: Plane trees
Platanus spp.
Woodland Garden

Below: Italian Alder
Alnus cordata
Woodland Garden

Bushy Park

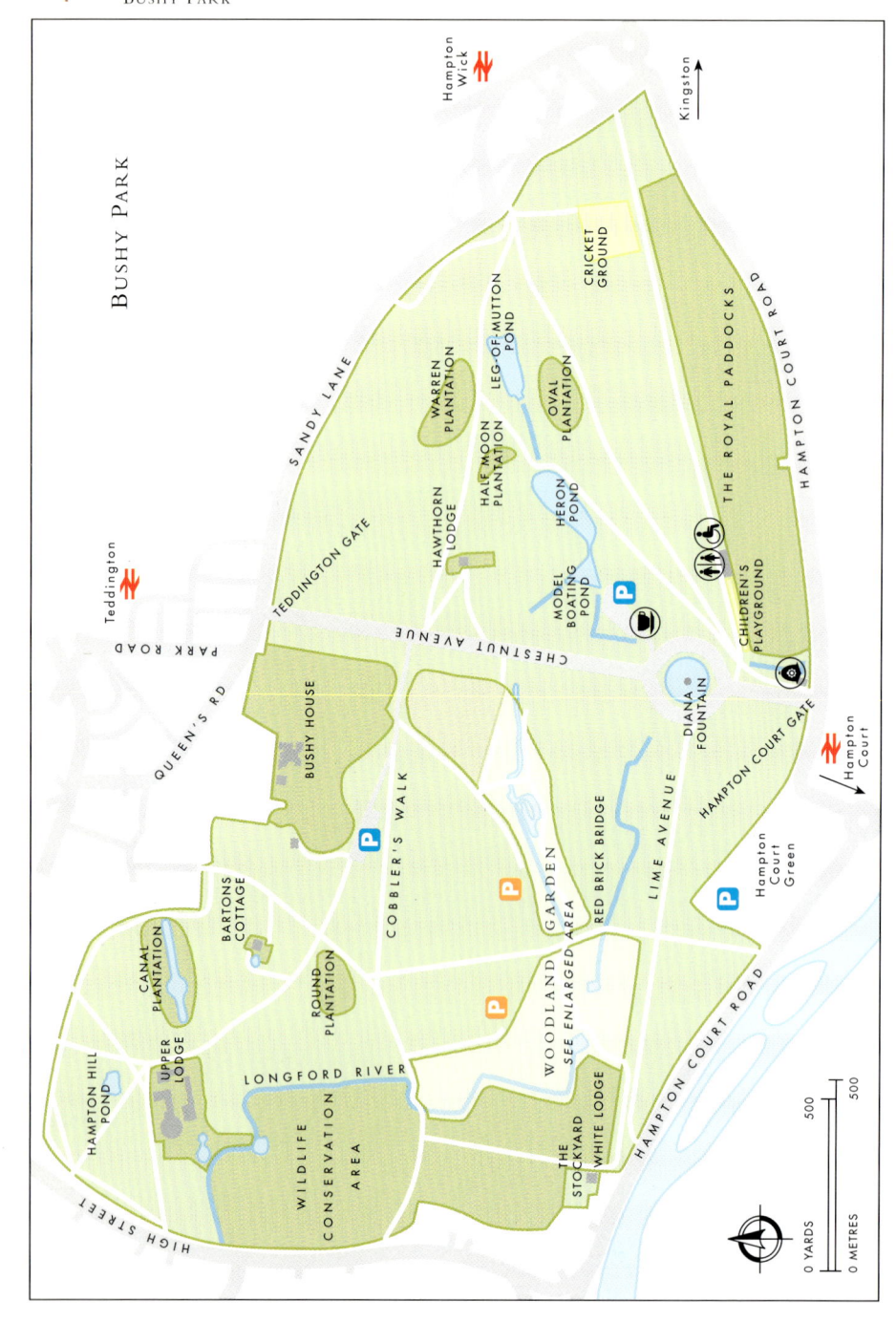

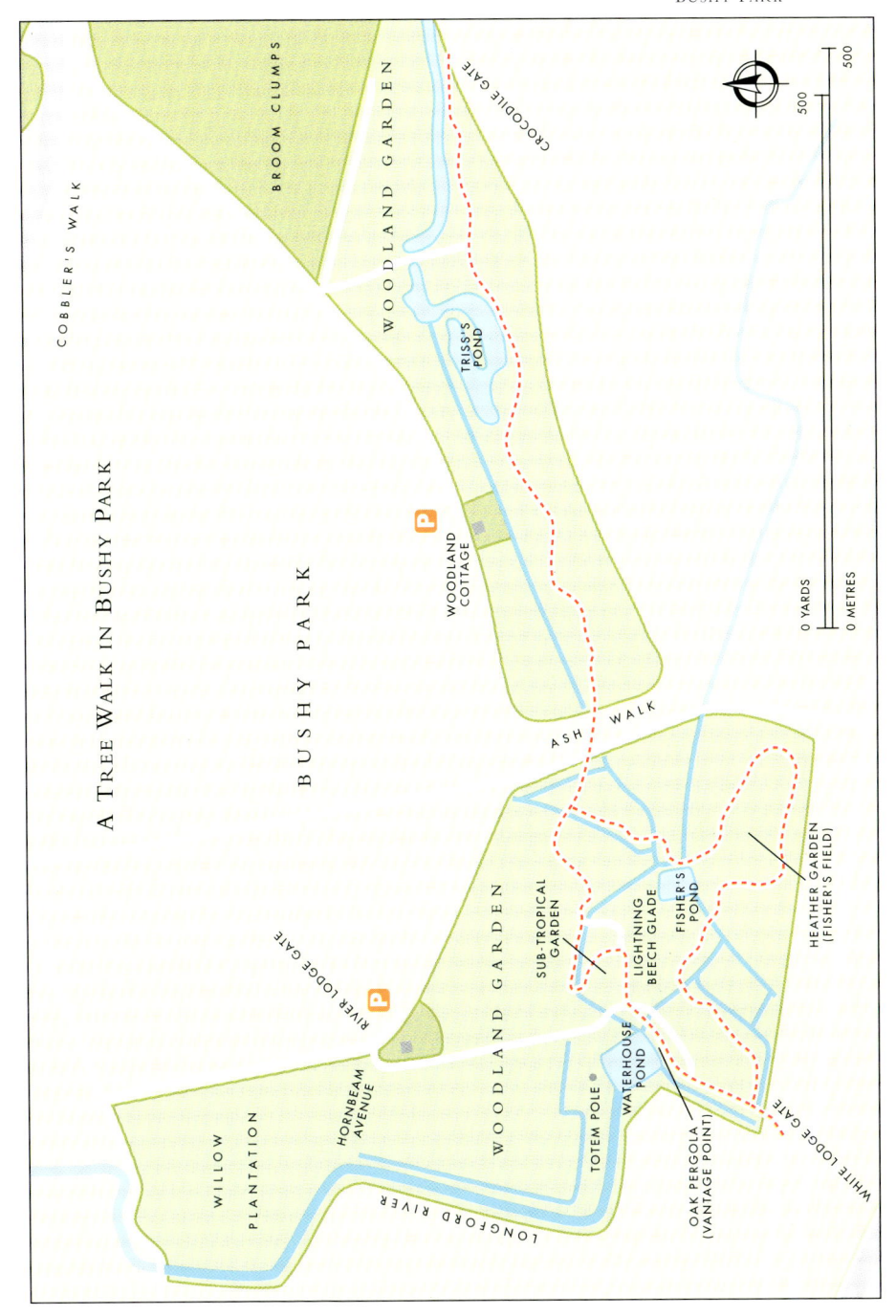

Envoi

THERE were elmes grete and stronge,
   Maples, ashe, ook, planes longe,
   Fyn ew, poplar, and lindes faire,
And other trees ful many a payre.
What sholde I tell you more of it?
There were so many trees yit,
That I sholde all encumbered be
Ere I had rekened every tree.

From Geoffrey Chaucer's translation
of *The Romaunt of the Rose.*

Glossary

Bottom-working Grafted or budded plants which have the scion inserted close to soil level, as opposed to **top-working** where the scion is inserted on a leg or stem 0.5–2m (20in–6ft) high.

Clone Vegetatively produced offspring of a single parent plant to which it is genetically identical.

Cultivar Plant arising and/or maintained in cultivation, which, when reproduced sexually or asexually retains its characteristics.

Gapping-up Filling gaps in planted areas that have been created by plants which fail to thrive.

Hybrid Plant produced by the cross-breeding of two or more genetically dissimilar parents.

Indumentum Covering of hair, scurf or scales; most often used in the general sense of 'hair'.

Patte d'oie Literally 'goose-foot'; a semi-circle of trees from which a number of avenues radiate in the manner of a bird's foot. Also referred to in Royal Parks and other ancient gardens as 'Great Bows' or 'Rounds'.

Pneumatophore Air-absorbing upward-growing root, functioning as a respiratory organ, found in plants of wet places.

Raceme Unbranched flower cluster with several or many stalked flowers borne singly along the main axis.

Scion Bud, or small portion of a plant (slip or graft) which is grafted on to the living root or rooted stem of a closely related plant.

Stock (Rootstock) Short, rooted stem which is chosen to support a bud or graft (scion).

Umbel Usually flat-topped or rounded flower cluster in which the individual flower stalks arise from a central point. In a compound umbel each primary stalk ends in an umbel.

BIBLIOGRAPHY

Addison, J. (1712) *The Spectator* No 477.
Bean, W.J. (1977–1980) *Trees and Shrubs Hardy in the British Isles.* (Supplement, 1988). John Murray, London.
Brickell, C. [ed.] (1989) *Royal Horticultural Society Gardeners' Encyclopedia of Plants and Flowers.* Dorling Kindersley, London.
Dendrophilus (1823) 'Discovery of the Secret Destroyers of the Trees in St James's Park.' *Philosophical Magazine* No 62, pp252–4.
Griffiths, M. (1994) *Royal Horticultural Society Index of Garden Plants.* Macmillan, London and Basingstoke.
Hillier, H.G. (1991) *The Hillier Manual of Trees and Shrubs.* 6th Edition, Hillier Nurseries (Winchester) Ltd.
Mitchell, A.F. (1982) *The Trees of Britain and Northern Europe.* Collins, London.

INDEX OF TREES IN THE ROYAL PARKS

Page references to illustrations are given in **bold**.

Alder (*Alnus* spp.) 12, 47, 84
   Common (*Alnus glutinosa*) 19, 30, 61, 83
   Italian (*Alnus cordata*) 83, **85**
Ash (*Fraxinus* spp.) **16**, 19, **46**, 79, 84
   Golden (*Fraxinus excelsior* 'Jaspidea') 20
   Narrow-leaved (*Fraxinus angustifolia*) 50
   One-leaved (*Fraxinus excelsior* 'Diversifolia') 61, **61**
   Shrew Ash 67
   Weeping (*Fraxinus excelsior* 'Pendula') 28, 32, **48**, 51
Australian Sassafras (*Atherosperma moschatum*) 72

Beech 20, 70, **73**, 79, 83
   Antarctic (*Nothofagus antarctica*) 62, 69
   Common (*Fagus sylvatica*) 56, 58, **60**, 69
   Copper **see** Beech, Purple
   Dawyck (*Fagus sylvatica* 'Dawyck') 40
   Fern-leaved (*Fagus sylvatica* 'Asplenifolia') 58
   Purple (*Fagus sylvatica* 'Purpurea') 18, 20, 39, 49
   Roble (*Nothofagus obliqua*) **80**, 82
   Weeping (*Fagus sylvatica* 'Pendula') **17**, 18, 20, **23**, 29, 31, **31**, 32, 38
Birch 20, 49
   Himalayan (*Betula utilis*) 18
   Jacquemont's (*Betula utilis* var. *jacquemontii*) **82**, 84
   Paper-bark (*Betula papyrifera*) **10**, **60**
   Red-barked (*Betula albosinensis* f. *septentrionalis*) 19
   Silver (*Betula pendula*) 18
   Swedish (*Betula pendula* 'Dalecarlica') 71
Bittersweet, Oriental (*Celastrus orbiculatus*) **8**, 82
Bladdernut Tree (*Staphylea colchica*) 59

Cabbage Tree (*Cordyline australis*) 85
Californian Laurel (*Umbellularia californica*) 31, 40, 61
Castor-oil Tree, Prickly (*Kalopanax pictus*) **56**, 61
Caucasian Elm (*Zelkova carpinifolia*) 17, 48, 49
Cedar 31, 58, 61, 80
   Atlas (*Cedrus atlantica*) 48
   Blue (*Cedrus atlantica* f. *glauca*) 49, 58, 81
   Lebanon (*Cedrus libani*) 31
Cherry
   Bird (*Prunus padus*) 70
   Tibetan (*Prunus serrula*) **35**, 41, 71

91

Wild (*Prunus avium*) 58, 70
Winter-flowering (*Prunus subhirtella* 'Autumnalis Rosea') 39, 84
Yoshino (*Prunus* x *yedoensis*) 48
Chestnut, Sweet (*Castanea sativa*) **6**, 8, 10, 19, 30, **55**, 56, 58, 60, **62**, 80
Chusan Palm (*Trachycarpus fortunei*) 32, 59, 85
Cornelian Cherry (*Cornus mas*) 18
Cypress, Lawson (*Chamaecyparis lawsoniana*) 69, 82, 84

Date Plum (*Diospyros lotus*) 28, **28**, 32
Deciduous Camellia (*Stuartia pseudocamellia*) 31, 70
Deodar (*Cedrus deodara*) 31, 48, 58, 61
Dogwood, Flowering (*Cornus florida*) 58, 62
Douglas Fir (*Pseudotsuga menziesii*) 50
Dove Tree (*Davidia involucrata*) 49, 59

Elm 37, 56
    Common (*Ulmus procera*) 36
    Japanese Golden (*Ulmus* 'Sapporo Autumn Gold') 19
Eucryphia
    Eucryphia glutinosa 83, 84
    Eucryphia x *intermedia* 82
    Eucryphia x *nymansensis* 'Nymansay' 84, 85
Euodia (*Tetradium daniellii*) 28, 41, 47

False Acacia (*Robinia pseudoacacia*) 36, 48, 58
    Golden (*Robinia pseudoacacia* 'Frisia') 40
    Pink-flowered (*Robinia* x *ambigua* 'Decaisneana') 18
    Single-leaved (*Robinia pseudoacacia* 'Unifoliola') 20
Fan Palm, Dwarf (*Chamaerops humilis*) 85
Fig 41, 59
    Common (*Ficus carica*) 40, **50**

Ginkgo (*Ginkgo biloba*) **9**, 17, 29, 31, 48, **59**, 61, 82
Golden Rain Tree (*Koelreuteria paniculata*) 41, 82
Gum (*Eucalyptus* spp.) 40
    Cider (*Eucalyptus gunnii*) 20, 48
    Snow (*Eucalyptus niphophila*) **79**, 83

Handkerchief Tree **see** Dove Tree
Headache Tree **see** Californian Laurel
Hickory, Shagbark (*Carya ovata*) 61
Holly 27, 49, 72
    Highclere (*Ilex* x *altaclerensis*) 17, 18, 57
Honey Locust, Golden (*Gleditsia triacanthos* 'Sunburst') 47, 48

Hornbeam 31, 60
    Common (*Carpinus betulus*) 47, **64**, 72
    Fastigiate (*Carpinus betulus* 'Fastigiata') **44**, 78, 79
Horse-chestnut (*Aesculus hippocastanum*) 17, 19, 20, 50, 56, 57, **76**, 77, **77, 84**
    Indian (*Aesculus indica*) 20, 27, 41

Indian Bean Tree (*Catalpa bignonioides*) 19, 20, 30, **36**, 48, 58, 62
    Golden (*Catalpa bignonioides* 'Aurea') **25**, 30
Ironwood, Persian (*Parrottia persica*) **7**, 32, 61, 83

Judas Tree (*Cercis siliquastrum*) 30, **38**, **39**, 40, 49, 50

Keaki (*Zelkova serrata*) 17, 29, 80
Kentucky Coffee Tree (*Gymnocladus dioicus*) 49

Laurel, Bay (*Laurus nobilis*) 20, 27, 29, 49, 72
Lime 19, 20, 27, 49, 77, 78, 79
    American (*Tilia americana*) 20
    Broad-leaved (*Tilia platyphyllos*) 26
    Caucasian (*Tilia* x *euchlora*) 30
    Common (*Tilia* x *europaea*) 29, **78**
    Pendent Silver (*Tilia* 'Petiolaris') 20, 40
    Silver (*Tilia tomentosa*) 26, 40, **41**
    Small-leaved (*Tilia cordata* 'Greenspire') **45**

Magnolia 59
    Chinese Evergreen (*Magnolia delavayi*) 19
    Common (*Magnolia* x *soulangeana*) 18, **54**
    Southern (*Magnolia grandiflora*) 12, 59, 72, 84
    Magnolia 'Kewensis' **79**, 83
Maple 39
    Ash-leaved (*Acer negundo*) 61
    Cappadocian (*Acer cappadocicum*) 49, 58, 59, 80
    Caucasian **see** Maple, Cappadocian
    Cut-leaved Silver (*Acer saccharinum* 'Wieri') 17
    Field (*Acer campestre*) 20
    Japanese (*Acer* spp.) 18, 59
    Montpelier (*Acer monspessulanum*) 30
    Nikko (*Acer nikoense*) 61
    Norway (*Acer platanoides*) 26, **30**, 50
    Paper-bark (*Acer griseum*) 19, **41**, 81
    Schwedler's (*Acer platanoides* 'Schwedleri') 12, 28, 50
    Silver (*Acer saccharinum*) 12, 48, 84
    Syrian (*Acer syriacum*) 31

Medlar (*Mespilus germanica*) **13**, **24**, 29, 30, 40, 84
Moosewood (*Acer pensylvanicum*) 59
Mulberry 61
   Black (*Morus nigra*) **15**, 17, 27, **29**, **36**, 38, 48
   White (*Morus alba*) 38

Oak 19, 60, 65, 66, 70, **70**, 72, 73, **73**, 79
   American 20, 49, 84
   Common (*Quercus robur*) 10, 50, 57, 60, **65**, **67**, 68, **71**, 80, 83
   Cork (*Quercus suber*) 20, 31
   Cypress (*Quercus robur* 'Fastigiata') **8**, 17, 18
   Daimyo (*Quercus dentata*) 82
   Holm (*Quercus ilex*) 20, 29, 31, 40, **47**, 49, **57**, 58, 80
   John Martin's Oak 8, 66, **67**, 73
   Lucombe (*Quercus* x *hispanica* 'Lucombeana') 17
   Pin (*Quercus palustris*) 20, **82**, 84
   Queen Elizabeth's Oak 60
   Red (*Quercus rubra*) **9**, 16, 28, 29, 40, 49, 58, 60
   Scarlet (*Quercus coccinea*) 40, **40**, 51, 58
   Turkey (*Quercus cerris*) 48, 50, 56, 58, 82
Osmanthus heterophyllus 32, **32**

Paulownia (*Paulownia tomentosa*) 27, **48**, 49
Pear (*Pyrus* spp.) 50
   Willow-leaved (*Pyrus salicifolia*) 51
Pine
   Austrian (*Pinus nigra* subsp. *nigra*) 29, 58, 69
   Bhutan (*Pinus wallichiana*) 61
   Big-cone (*Pinus coulteri*) 81
   Monterey (*Pinus radiata*) 80, 81
   Scots (*Pinus sylvestris*) 61, 81, 83
   Weymouth (*Pinus strobus*) 81
Pittosporum (*Pittosporum tenuifolium*) 17, 61
Plane (*Platanus* spp.) **11**, **14**, 19, 32, 37, 39, 47, 56, **85**
   American **see** Plane, Western
   London (*Platanus* x *acerifolia*) 10, 11, 12, 17, 29, **34**, **37**, 39, 80
   London (cultivar) (*Platanus* x *acerifolia* 'Augustine Henry') 12, 39, 40
   Oriental (*Platanus orientalis*) 10, 11, 58
   Western (*Platanus occidentalis*) 10, 11
Poplar 39, 83
   Balsam (hybrid) (*Populus balsamifera* x *trichocarpa* Clone 32) 47
   Black (*Populus nigra*) 12, 30
   Grey (*Populus canescens*) 41, 49
   Lombardy (*Populus nigra* 'Italica') 12, 41, 47, 50
Pride of India **see** Golden Rain Tree
Privet, Tree (*Ligustrum lucidum*) 20, 49
   Variegated (*Ligustrum lucidum* 'Excelsum Superbum') 30, 58, 59

Red Cedar, Western (*Thuja plicata*) 84
Redwood
    Coastal (*Sequoia sempervirens*) 72, 80
    Dawn (*Metasequoia glyptostroboides*) 12, 18, 29, 48, 62, 69, 71, 84
Reformers' Tree 16, 20

Sequoia, Giant (*Sequoiadendron giganteum*) 72
Snowbell Tree (*Styrax japonica*) **69**, 70
Snowdrop Tree (*Halesia carolina*) 71
Spindle (*Euonymous* spp.) 82
    Euonymus myrianthus 69
Spruce
    Colorado Blue (*Picea pungens* var. *glauca*) 84
    Serbian (*Picea omorika*) 81
Swamp Cypress (*Taxodium distichum*) 12, 30, 38, 39, 50, 51, 62, 68, 80, 81, **81**, 83
Sweet Gum (*Liquidambar styraciflua*) 26, 58
Sycamore (*Acer pseudoplatanus*) 20, 57

Tree of Heaven (*Ailanthus altissima*) 38, **39**, 49, 83
    Downy (*Ailanthus vilmoriniana*) 83
Tulip Tree (*Liriodendron tulipifera*) 27, **27**, 49, 57, 61, 69
Tupelo (*Nyssa sylvatica*) 61, 71

Varnish Tree, Chinese (*Rhus potaninii*) 20

Walnut 20, 61
    Black (*Juglans nigra*) 19, **19**, 27, 50
    Common (*Juglans regia*) 19, 38
Willow 12
    Babylon (*Salix* x *sepulcralis*) 12, 30, 41, 50, 51, **51**, 68
    Corkscrew (*Salix matsudana* 'Tortuosa') 48
    Crack (*Salix fragilis*) 12, 47, 85
    Golden Weeping (*Salix* 'Chrysocoma') 12, 30, 41, 51, 68, **83**
    Silver (*Salix alba* f. *argentea*) 12, 47
    White (*Salix alba*) 12, 19, 30, 41, 47, 51, 62, 68
Wingnut 20
    Caucasian (*Pterocarya fraxinifolia*) 18, **18**, 20, 80
    Hybrid (*Pterocarya* x *rehderiana*) **15**, 17

Yellow Wood, Kentucky (*Cladrastis lutea*) 17
Yew, Common (*Taxus baccata*) **22**, 29, 31, 49

95

THE
ROYAL
PARKS